Revision Exercises in Basic Engineering Mechanics

to test your knowledge and for revision

Gregory Pastoll, PhD (Higher Education) BSc Mech Eng

Copyright and origination

This work was first published by Gregory Pastoll in 2023.

Cover artwork by Gregory Pastoll

Typeset by Book Covers Australia
bookcoversaustralia.com

ISBN   978-0-6452688-8-1   (paperback)

ISBN   978-0-6452688-9-8   (e-book)

Contents

The purpose of this book

This book presents a variety of exercises in basic engineering mechanics, for testing your own knowledge.

These exercises are applicable to the study of this subject at university, college and high school level. The topics match the core of most curricula in basic engineering mechanics.

Students will find the exercises helpful:

- To build their understanding of the subject, and
- To prepare for being assessed.

Instructors are welcome to use or adapt any of the exercises for their own use in teaching and assessment.

All the questions and the illustrations have been compiled by the author, and may be used, adapted or copied without seeking permission from the publisher, provided the source is acknowledged.

All calculation questions are supplied with answers.

Certain questions designed for tackling in groups are not supplied with answers, as the outcome of such exercises depends on the approach taken by participants.

Note: the methods of solving these exercises are not given here.

All the principles and methods of solving exercises in basic mechanics can be found in the three volumes of this author's series of books that is described in the Appendix to the present book.

Anyone who can answer all the questions in the present book correctly, definitely has a thorough knowledge of the principles of mechanics, and is ready to apply those principles in their further study of engineering.

Getting the most out of testing your own knowledge

Your learning is not served by trying to absorb someone else's knowledge structure, but by building your own.

When you study a formal course in any subject, you will most likely get assessed. It is easy to get the impression that your aim is to be able to score good grades by reproducing what your assessors seem to want.

If that *remains* your personal aim, you will find it unsatisfactory in the long run. Sure, you need to pass, but more importantly, you need to understand the course content well enough, so that you can apply it in your working career.

The number one requirement for building a sound knowledge structure is that you should be absolutely honest with yourself. If you don't understand some principle, be bold enough to admit it to yourself. That is the starting point for developing a proper understanding.

People who *think* they understand something, but don't fully understand it, are deluding themselves. Let's hope, dear reader, that you are not one of them.

A person's delusion index

This index describes the extent to which a person is deluding themselves about the meaning of certain words and concepts. It is measured using the 'word-misconception test' that the present author helped to develop.

This test was issued to first-year students of physics at a university. It was designed to determine the extent to which entering students whose home language was English understood the meaning of a set of commonly-used words encountered in high school physics.

The words we tested were simple enough, if you had done physics at school. The test covered 25 words, including, for example:

'random', 'density', 'point', 'pressure', 'theory', 'device', 'technology' and 'vertical'.

The test had two stages. In stage 1, respondents were asked to state whether they knew the meaning of each respective word. They had three options: 'yes, I understand this word'; 'I have some idea of what it means'; and 'I don't know what this word means'.

Their quiz papers were then taken in. In stage 2 of the test, they were again presented with the same list of words, only, this time, with each word, there was list of five possible meanings. A repondent had to say for each of these possible meanings, whether it was correct, or not. Some of the meanings were correct, and some not. In order to be considered to fully understand the meaning of any one of the given words, you had to correctly identify the applicability of *all five* given meanings. If you got the whole pattern correct, you knew the meaning of the given word.

Example: the word **'feasible'** (indicate Y = yes, this is a correct meaning, or N = no, this meaning is not correct)

Possible definition		
Can be done and is economically worthwhile	Y	N
Can be done, irrespective of cost	Y	N
Will not harm the environment	Y	N
Is economically attractive and has backers	Y	N
Will make a huge profit	Y	N

If you incorrectly identified the applicability one or more of the meanings provided, you were considered *not* to know what the word meant.

If you said in stage 1 that you knew what the word meant, but it turned out from your answers to stage 2 that you didn't actually know, then you were considered to be deluded about your knowledge of that word. Your delusion index was the number of words you were deluded about, out of the 25 words in the test.

No-one would be too concerned if they tripped up on one or two words out of 25. The shocking result, though, was that, of a large sample of students tested, the *average* delusion index was 15 out of 25, namely 60%.

If you want to read up the paper on the test that was done, this is the reference: Jacobs, Glenda (1989) 'Word usage misconceptions among first-year university physics students', Int.J.Sci.Educ. Vol 11 no.4, 395 - 399.

The results of that test bring us all face-to-face with a choice. Are we prepared to sail through life (and college) with only a vague understanding of the terms and concepts we are going to encounter every day in our profession? If we are, how can we expect to be regarded as responsible or reliable? This question is particularly important for engineers, because engineering decisions simply cannot be left to vagueness or guesswork. The way that the physical world behaves won't let them.

A true/false test can give you the means to identify what you don't know.

To assist you to develop that self-honesty, over forty true/false tests are supplied in this book. When you tackle any such test, there is a particular scoring system you should apply, to make sure that you *are* being honest with yourself, and to identify the points about which you need to seek clarification.

Here is an example of the structure of a true/false quiz on some randomly chosen topics in basic engineering mechanics. Give it a quick try if you like. No-one is going to assess you on it. In fact, the tests that appear throughout this book are meant for you to mark yourself, not to get a grade, but to discover where you need clarification.

The '?' symbol is circled to admit you don't know whether the statement is true or not.

1	There are more than two states of physical equilibrium.	T	F	?
2	Momentum and inertia are entirely different concepts.	T	F	?
3	The value of the friction coefficient between two unlubricated given solids can exceed 1.0 in certain circumstances.	T	F	?
4	Newton's third law is about the acceleration of a mass by a force.	T	F	?
5	The centre of percussion of a rod of uniform section that is pivoted at one end, is found halfway down the rod.	T	F	?
6	The mechanical advantage of a simple lifting machine can sometimes be greater than the velocity ratio of that machine.	T	F	?

Clearly, the questions in this sample test cannot be addressed by calculations. They require knowledge and understanding. If you don't have the requisite grasp of the facts and principles, no matter how much mathematics you know, you won't be capable of doing engineering.

The commonly-used method of scoring a true/false test is to award one mark for each question you get correct, and zero for each one that you get wrong or omit to answer. However, by using that scoring method, you might 'pass' such a test merely by random guesswork. So, that scoring method doesn't lead to any self-examination. The following scoring method *does*:

Suggested scoring for true/false tests:

- Each correct answer earns you two marks.
- If you say you don't know the answer, your honesty is rewarded by being allocated one mark.
- If you omit to answer, nothing can be deduced about your state of knowledge. You might not understand the question, or you might understand it but have no clue about how to answer it. Maybe you ran out of time and didn't get to that question. Or, you might be leaving it until later to come back to it and have another think. Either way, for omitting to answer, you are awarded zero.

- Incorrect answers are penalised, to discourage guessing. Why? Because in engineering, it is unacceptable to guess. A wrong guess on the job could result in damage to property, failure of design, waste of resources and danger to personnel. For each answer that you get wrong, 3 marks are deducted.

With this scoring system, for a test of 10 questions:

Supposing you got them all correct, you would earn a score of 20.
If you admit to not knowing *any* of the answers, you would score 10.
If you guess them all and get half of them correct by chance, and the other half wrong, you would score –5.
If you knew half of the answers and admitted to not knowing the rest, you would score 15.
A pass-mark could be nominated as any score between 15 and 20.

The point of admitting you don't know is to establish exactly *which* questions are giving you trouble, so you can take steps to find out how to improve your knowledge.

If such a test were issued as an exercise by an instructor, with access to the students' responses, it would immediately be clear which questions the class as a whole admitted to be unable to answer. Steps could then be taken to remedy the situation by focusing additional explanations on the areas that were not well-understood.

The way each set of exercises in this book is organised

The topics under which the exercises are grouped are the same as those used in the author's three-volume series of books: 'Basic Engineering Mechanics Explained', described in the Appendix .

In each set are included, relevant to that topic, some or all of the following:

- True/false tests
- Short questions requiring descriptive answers, and
- Calculation questions

The answers to the true/false tests are given at the back of the book.

The answers to the calculation exercises are provided directly after each exercise.

Answers are *not* supplied for the questions that require short descriptive answers, for three reasons:

- If unsure of the best way of answering these questions, it will serve the reader better to consult peers and instructors, so that you can compile your own answer, than to have the solution presented on a plate.

- All of them can be answered by reading the explanations in the author's series 'Basic Engineering Mechanics Explained', and

- If the present book were to provide such descriptive answers, even one paragraph for each question, it would make the book far too long.

Being able to answer this type of short question is extremely important, because the best indication of your grasp of a principle is how well you can explain it to someone who is not trained in your discipline.

1

Concepts, quantities, principles and laws of mechanics

Archimedes

True/false tests on this topic

	True/false Test # 1a Concepts, quantities, principles and laws of mechanics			
1	There are only two units of mass used throughout the world.	T	F	?
2	Historical measures of length were often based on typical dimensions of the human body.	T	F	?
3	All units of measure have been chosen arbitrarily.	T	F	?
4	There are three fundamental quantities in Mechanics.	T	F	?
5	A 'radian' is not based on any of the three fundamental units because it is defined as a ratio of one length dimension to another.	T	F	?
6	A weighing scale actually measures mass, not weight.	T	F	?
7	The weight of one kilogram is known as a kilogram force.	T	F	?
8	The weight of of an object with one kilogram of mass will be identical no matter where the object is weighed.	T	F	?
9	A force moment has the same units as a torque.	T	F	?
10	In the SI system, the unit of energy is the calorie.	T	F	?

	True/false Test # 1b Concepts, quantities, principles and laws of mechanics			
1	The length of a metre is currently defined by the length of a particular platinum bar that was designed to be one ten-millionth of the distance along the surface of the Earth from the north pole to the equator.	T	F	?
2	The basic unit of mass in the SI system is the gram.	T	F	?
3	Force is measured in newtons.	T	F	?
4	The weight of one tonne is approximately 1000 newtons.	T	F	?
5	A typical grown man could weigh approximately 80 N.	T	F	?
6	The definition of the magnitude of a newton is derived by applying Newton's Second Law.	T	F	?
7	A healthy, fit student could easily exert a horizontal force of 600 N during a tug-of-war.	T	F	?
8	Mass is a concept, rather than a measurable quantity.	T	F	?
9	The Parallelogram Law for the addition of vectors applies to all types of vector in two dimensions.	T	F	?
10	Archimedes measured the volume of an irregular object by determining how much water it displaced.	T	F	?

Questions requiring short descriptive answers

1. Can Newton's law of gravitation be applied to determine the gravitational force between two objects, when one of them is inside the boundary of the other, and exactly in the centre, like a screw inside a nut? Explain.

2. Are all units of length chosen arbitrarily? Why, or why not?

3. How can you measure the mass of an object without weighing it?

4. Under what circumstances is the velocity of an object identical with its speed?

5. If the Newtonian view of the physical universe has been made obsolete by the scientific theories of Einstein and others in the 20th century, why is it still used by engineers?

6. The measurement of which quantity has been altered to conform with a revised measurement of the speed of light in a vacuum?

7. How was the length of a metre initially defined?

8. The units of force are defined by the units of which other quantities, and in what relationship?

9. How does it help our calculations to assume that a body acted on by forces remains 'rigid'?

10. Name three machines in current use which are entirely mechanical, in the sense that they do not make use of heat or electricity.

11. Name and describe five of the fundamental principles or 'laws' that govern the science of Mechanics.

12. Name five derived quantities used in Mechanics and state the units of each, in terms of the units of the fundamental quantities.

13. Name five different sources of force, besides muscular effort and magnetism.

14. Why is the kilogram, and not the gram, considered to be the fundamental unit of mass?

15. Explain why it is necessary to understand basic mechanics when designing a robot.

16. Define what is meant by a 'particle'.

17. Is there a difference between a 'principle' and a 'law' applying to the science of mechanics? Explain.

18. What distinguishes the three 'fundamental' quantities from any other quantities used in the science of mechanics?

19. Describe the zeroth law of mechanics.

20. Can there really be such a thing as a 'closed system'? Explain.

21. Give one example of a machine that is used to transform an applied physical force into a force exerted in a different direction.

22. Explain why a 'factor of safety' is sometimes called a 'factor of ignorance'.

23. State Newton's first law and describe one application of this law.

24. State Newton's second Law, and describe an example of where this is observed in practice.

25. State Newton's third Law, and provide an example of it.

26. If it were possible to get yourself to the centre of the Earth, and temperature was no consideration, then, according to Newton's Law of Gravitation, your weight would be infinite. Do you agree? Explain.

27. What does Archimedes' Principle enable us to do?

28. Define the Law of Conservation of Energy.

29. What does the Law of Conservation of Momentum state?

30. Suppose you have a solid copper ball and a hollow aluminium sphere of the same size. If you drop them both from a high balcony at the same instant, will they hit the ground together, or will one of them hit first? Explain.

31. If you weigh a bucket of sand on a bathroom scale at sea level, and then take that same scale with you by car to the top of a high mountain pass, and weigh the bucket again, would the two readings be the same? Explain.

32. You are given a basket of stones and tasked with throwing them as far as possible over level ground. To standardise the effect of air resistance, all the stones are spherical, and the same size. However, they are of different densities. Which considerations affect how far a given stone can be thrown? Which of the principles of mechanics apply here?

Calculation exercises on quantities, principles and laws of mechanics

Exercise 1.a

If an astronaut weighs 800 N on the surface of the Earth, how much would he weigh when in a craft that is stationary relative to the Earth, and one Earth radius above the surface? [200 N]

Exercise 1.b

If the gravitational acceleration on the Moon is 1/6 that of Earth, and an astronaut in a space-suit on Earth can do a standing jump to clear a bar 400 mm off the ground: show by calculation what height she should be able to jump in that way on the moon. [2.4 m]

Exercise 1.c

Use Newton's Law of Gravitation to determine the gravitational acceleration on the surface of the moon, given that the mass of the moon is 1/81 that of the Earth, and their radii are 1738 km and 6371 km respectively. [1.6274 m/s^2]

Exercise 1.d

If the mass of a spacecraft is 20 tonnes, what thrust would its engines have to develop to enable it to change its speed from 180 km/h to 2160 km/h without changing direction, in a time of 20 minutes, and how far would it have travelled in that time? [9167 N and 390 km]

2

Expressing numbers in engineering

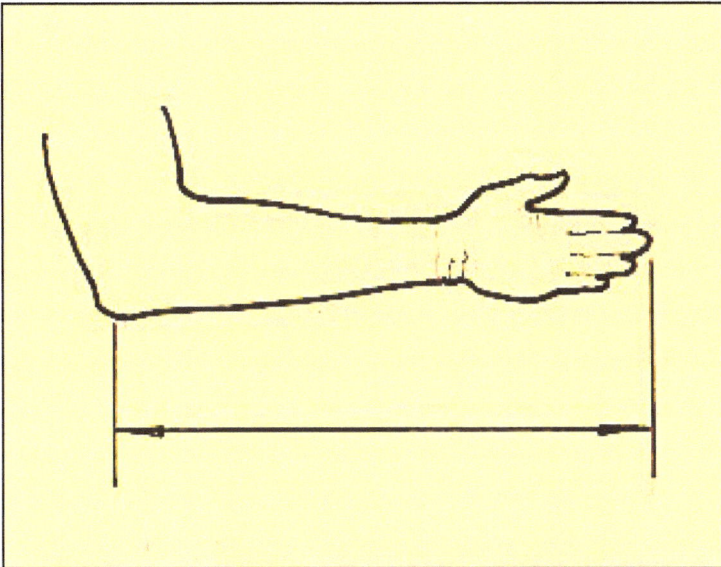

True/false tests on this topic

	True/false Test # 2a Expressing numbers in engineering			
1	In representing numbers, there is a difference between scientific notation and engineering notation.	T	F	?
2	Scientific notation requires only one digit before the decimal point.	T	F	?
3	Engineering notation should be used by engineers in all circumstances, preferably to ordinary arithmetic notation.	T	F	?
4	Ordinary arithmetic notation can be used instead of engineering notation in some circumstances.	T	F	?
5	A dimension of 4532 m would be written as 4.532×10^3 m in engineering notation.	T	F	?
6	The reason for engineering notation differing from scientific notation is that the former is more practical from the point of view of visualising quantities.	T	F	?
7	The number 0.0087032 would be written as 8.703×10^3 in engineering notation.	T	F	?
8	The number 5.0076×10^4 would be written as 0.5008×10^3 in engineering notation.	T	F	?
9	A 'significant digit' is a digit whose value helps to define the magnitude of a number for a particular purpose.	T	F	?
10	A 'significant digit' is any non-zero digit.	T	F	?

	True/false Test # 2b Expressing numbers in engineering			
1	Significant digits are those that appear after the decimal point in a number.	T	F	?
2	If a census result shows the number of citizens residing in a country to be 24 900 083, there are eight significant digits in this number.	T	F	?
3	All non-zero digits in a number are significant.	T	F	?
4	Some zeroes in a number can be significant.	T	F	?
5	In any number, after the decimal point, the zeroes to the left of the first non-zero digit are all significant.	T	F	?
6	If a weighing scale is accurate to within one tenth of a gram, and it records the weight of an item as 652.0 g, then this figure has four significant digits.	T	F	?
7	In most instances in engineering, presenting a number correct to four significant figures is accurate enough for all practical purposes.	T	F	?
8	In an answer that results from a calculation, the more digits there are after the decimal point, the more accurate the calculation.	T	F	?
9	The number 2.4000765×10^7, when converted to engineering notation, should read 24.001×10^6.	T	F	?
10	During a complex calculation, if all intermediate answers are rounded off to four significant figures, the accuracy of the end result would not be significantly different from the true final answer.	T	F	?

Questions requiring short descriptive answers:

1. Explain why engineering notation is more practical for engineers than is scientific notation.
2. Show by example the difference between the accuracy and the precision of a set of measurements.
3. In a calculation in which five numbers have to be manipulated as follows: $a + b[c \div (d - e)]$: if you suspect there to be a slight inaccuracy in the value of one or more of the numbers, which part of the calculation would have the most disruptive effect on the accuracy of the final answer? Explain.
4. Give an example of a number that looks different in ordinary arithmetic notation and engineering notation.

Calculation exercises:

Exercise 2.a

This is a unit-conversion exercise. It should be done showing all the necessary steps, without using conversion tables found in a calculator or computer.

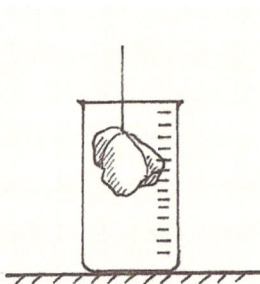

An irregularly shaped stone is suspended from a light thread and weighed in air. Result: 2 lb 3 ounces.

The stone is then dipped into a measuring cylinder which is full to the brim with water. When the stone is taken out of the water, the volume of water in the cylinder has diminished by 7 fluid ounces.

Given that: 1 lb = 16 oz; 1 pint = 16 fluid ounces; 8 pints = 1 gallon; 1 lb = 454 grams; and 1 gallon = 4.53 litres:

Determine the density of the stone in kg/m³. [4009 kg/m³]

Exercise 2.b (a unit-conversion exercise)

A meteorite found in a desert crater is roughly spherical. Its diameter is reported to be almost exactly four feet.

Given access to tables of density and the fact that that 1 inch is exactly equal to 2.54 cm:

If its mass was found to be 9.4 tonnes, could it have consisted of pure iron? [no, its density = 8052 kg/m^3 which is significantly greater than that of pure iron, so it must contain some other heavier metals.]

Exercise 2.c (a unit-conversion exercise)

Given that 1 gallon (US) = 3.7854 litres and one foot = 30.48 cm:

Without referring to any unit-conversion tables, determine what a fuel consumption of 40 miles per gallon in US units would be in in units of km/litre. Show all the steps in your calculations. [17.01 km/l]

Exercise 2.d

Express all of the following numbers in engineering notation, correct to four significant figures:

a. 8504.8732 kg.......................[8.505 × 10^3 kg]
b. 122 349 Pa ....[122.3 × 10^3 Pa]
c. 218.463 m ....[218.5 m]
d. 0.0042093 N ....[4.209 × 10^{-3} N]
e. 0.0100403 km^2 ....[10.04 × 10^{-3} km^2]
f. 7.00452 × 10^5 light years.....[0.7005 ×10^6 light years]

3

Forces, components, resultants and equilibrium of particles

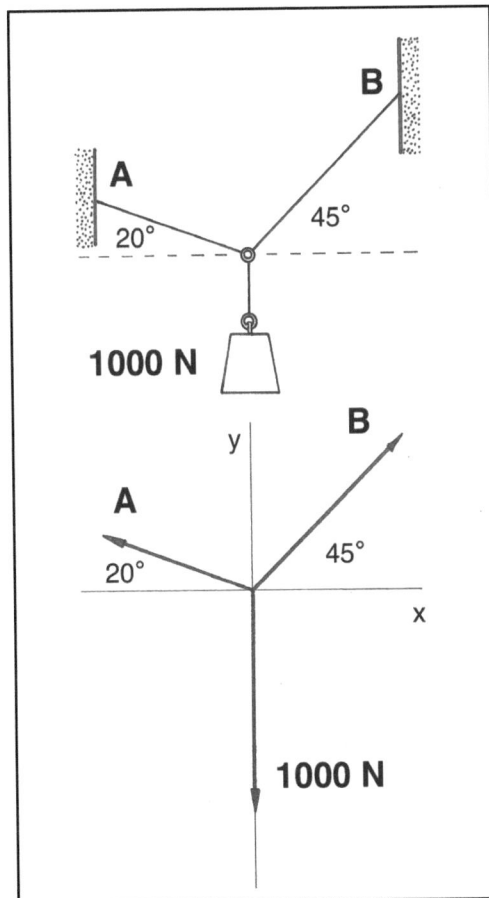

True/false tests on this topic

	True/false Test # 3a Forces, components, resultants and equilibrium of particles			
1	A force can only exist if it is opposed.	T	F	?
2	If a speeding train hits a mosquito, the forces they exert on one another are equal and opposite.	T	F	?
3	The average able-bodied adult person would easily be capable of exerting a force of 120 N by pulling on the ends of a spring-scale.	T	F	?
4	An electrical current could be used to give rise to a force.	T	F	?
5	There are less than five separate phenomena that can give rise to forces.	T	F	?
6	A weighing scale measures the amount of mass contained in an object.	T	F	?
7	If you determine the mass of a given brick on a particular electronic scale, the readout would be the same, whether the scale was at sea level or at the top of a high mountain.	T	F	?
8	Newton's Law of Gravitation, applied to determine the gravitational force between two bodies, holds, even when one of the two objects is physically inside the boundary of the other object.	T	F	?
9	The resultant of two vectors can never be obtained by adding their magnitudes arithmetically.	T	F	?
10	'Displacement' is a vector quantity.	T	F	?

	True/false Test # 3b Forces, components, resultants and equilibrium of particles			
1	The resultant of a set of forces acting together can only be determined if those forces occur in the same plane.	T	F	?
2	The resultant of a set of co-planar forces can be determined by any one of three different methods.	T	F	?
3	A given force cannot be represented by more than two components.	T	F	?
4	When a given force acts on an object, it will have exactly the same effect on the object irrespective of where it is applied, as long as it remains acting in the same line of action, in its original direction.	T	F	?
5	The first step in determining the direction or magnitude of an unknown force among a set of forces that are in equilibrium, requires a free-body diagram to be drawn of a point that is in equilibrium due to the action of those forces.	T	F	?
6	When analysing coplanar forces in equilibrium to determine the magnitude or direction of an unknown force, the method of using rectangular components is, in all circumstances, simpler to use than a graphical method.	T	F	?
7	If a set of forces is not in equilibrium, its resultant can still be determined.	T	F	?
8	If the forces acting on an object are in equilibrium, the object has to be standing still.	T	F	?
9	When a crane is used to raise a long beam, by engaging the crane hook with a wire rope sling attached to the beam at both ends of the beam: using a sling that is twice the length of the beam would be safer than using one that is four times the length of the beam.	T	F	?
10	If a child holds the string to which a helium balloon is attached, the force which the child exerts on the string can be described as the equilibrant to the other forces acting on the balloon.	T	F	?

Questions requiring short descriptive answers

1. Define what is meant by the 'components' of a vector.
2. How many different components can a vector have? Explain.
3. If you try to push a crate on a horizontal concrete floor and it doesn't move, what could you deduce about the magnitude of the friction force between the crate and the floor?
4. Under what circumstances can an object be said to be in equilibrium? Explain.
5. Describe two ways of specifying the direction of a 2-D vector.
6. Can a force have three or more components in the same plane? Explain with the use of a diagram.
7. What exactly is a 'force couple'? Explain with a sketch.
8. What effect does a force couple have on an object to which it is applied?
9. Give a clear definition of an 'equilibrant', as applied to a set of forces acting at a point.
10. What would be the effect on an object acted on by a set of forces, if an equilibrant was added to the set?
11. If a speeding car collides with a beach ball kicked across the road, are the forces that the two objects exert on one another equal in magnitude? Why, or why not?

Calculation exercises

Exercise 3.a

The figure shows four co-planar forces acting at point O.

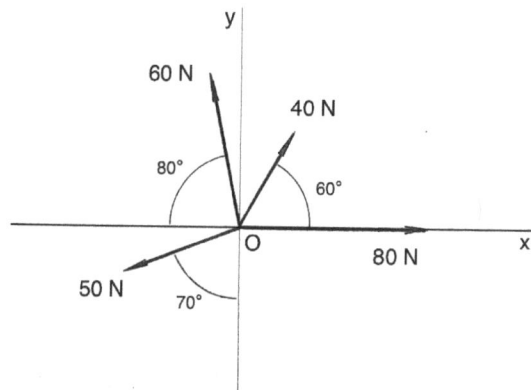

- Use a graphical method to determine the magnitude and direction of the equilibrant of these four forces. [the value obtained depends on the accuracy of your workings]
- Again, determine the magnitude and direction of the equilibrant, this time by summing up components. [87.67 N]
- Determine the percentage error for the value you obtained graphically. Show your reasoning for this calculation.

Exercise 3.b

Three co-planar forces acting at a point are in equilibrium.
Using a graphical construction, determine the magnitudes of force F and angle θ.
[110 N; 16°]

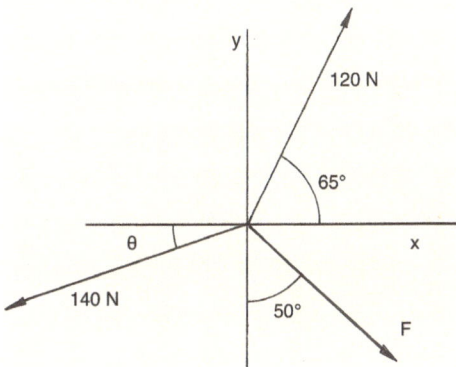

Exercise 3.c

This set of coplanar forces is in equilibrium. Construct to scale a polygon of forces, to determine the value of force F and angle θ.

[A graphical solution shows there are two possibilities: the analytical values for these are 76.89 N and 40.37°; 135.2 N and 9.63°]

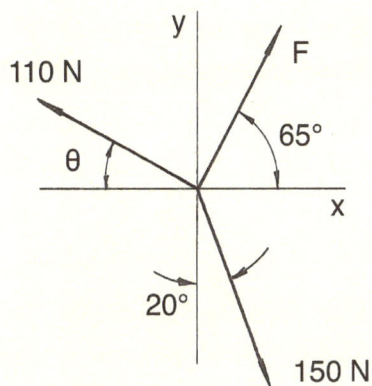

Exercise 3.d

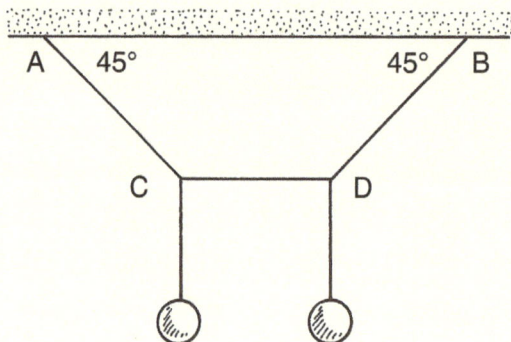

Two identical mass-pieces are suspended from chains that are anchored at points **A** and **B**.

A short chain, **CD**, is attached to these hanging chains, pulling them towards each other as shown.

By what percentage has the addition of the tie **CD** increased the pull on the attachments at **A** and **B**? [41.41%]

Exercise 3.e

A brick weighing 30 N is held in equilibrium against a frictionless ceiling by the two forces shown.

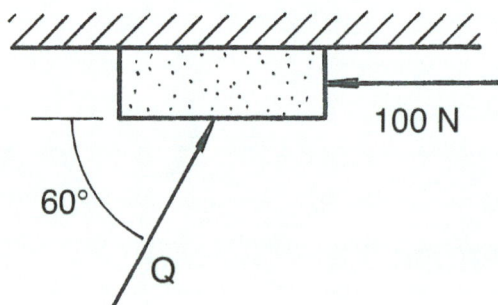

Determine:

- The magnitude of force **Q** [200 N] and

- The magnitude of force **N**, the normal reaction that the ceiling exerts on the brick. [143.2 N]

Exercise 3.f

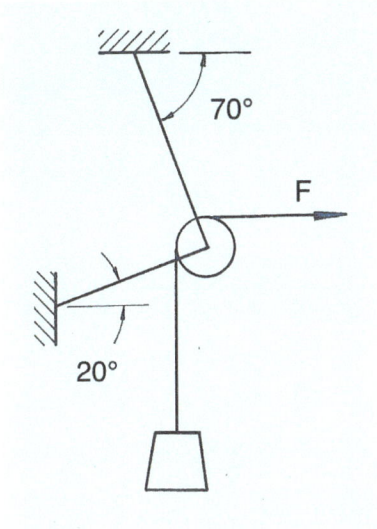

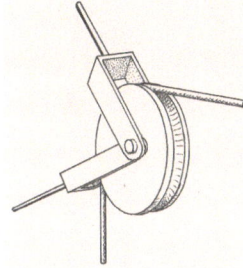

A sheave that is free to rotate with minimal friction is attached through its axle to two rods, each of which is hinged at its other end. A rope passing over the sheave is used to support a weight of 1000 N.

If it is in equilibrium in the position shown, determine the tensions in the two rods. [1282 N; 598 N]

Exercise 3.g

A vertical wooden pole is held up by two wires, pulling it down against a rough concrete floor. Assume the pole is stable in the plane of the paper.

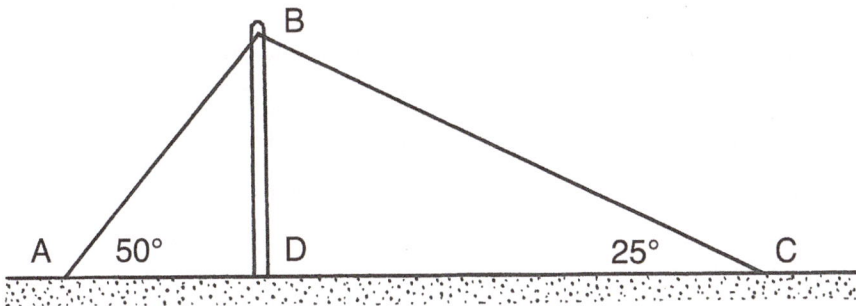

If the tension in wire **AB** is 1200 N, determine the tension in wire **BC**, and the compression in the pole. [851 N; 1279 N]

4

Force moments, torque and equilibrium of rigid bodies

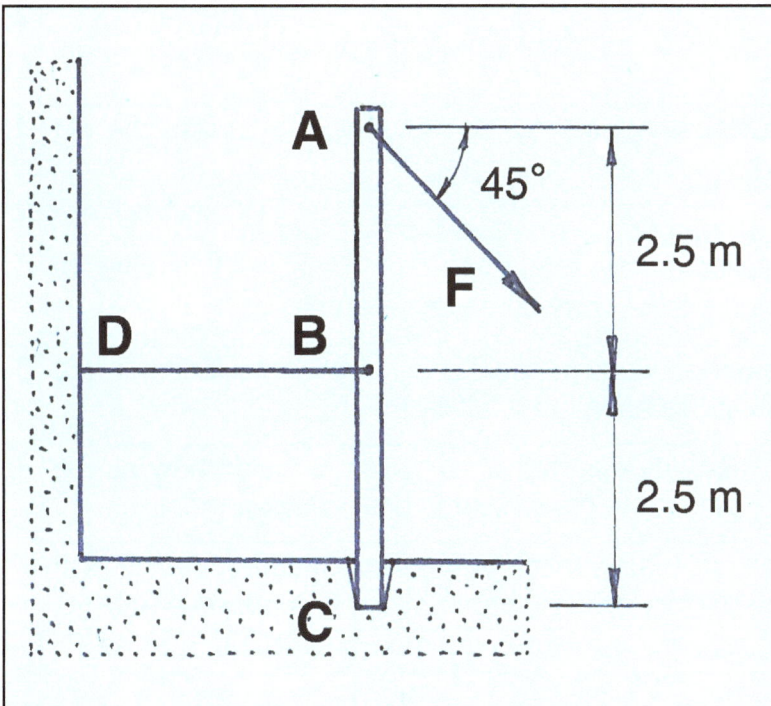

True/false test on this topic

	True/false Test # 4a Force moments, torque and equilibrium of rigid bodies	T	F	?
1	The value of a force moment in a given plane has to be specified in relation to a nominated axis of rotation.	T	F	?
2	The moment of a force is the rotating effect provided by that force.	T	F	?
3	A force moment is not a vector quantity.	T	F	?
4	The magnitude of the moment of a force couple is given by 2Fd, where 'F' is the magnitude of each of the forces and 'd' is the shortest distance between their lines of action.	T	F	?
5	A torque produces an equivalent effect to that of a force moment.	T	F	?
6	A solid object cannot be in equilibrium unless the net force moment on it is zero.	T	F	?
7	If a number of forces act on a rigid body that is in equilibrium, the total force moment about two different points on the body could be different.	T	F	?
8	For a rigid body in one plane to be in equilibrium, three conditions have to be fulfilled.	T	F	?
9	A free-body diagram of an object should show all the forces that the object exerts on its environment.	T	F	?
10	For a rigid body that is in equilibrium under the action of several forces in the same plane, it is possible to solve for as many unknowns as there are forces.	T	F	?

Questions requiring short descriptive answers

1. Name the conditions for the equilibrium of a rigid body in one plane.

2. What is meant by the term 'over-constrained' applied to a rigid object that is acted on by a number of forces?

3. Describe the number of unknown forces that need to be solved for at the support of a cantilevered beam.

4. Illustrate the physical appearances of a simple support on a loaded beam; and one that constrains the beam in the x- and y- directions.

5. How is a torque actually different from a force moment that is applied by a single force acting some distance from a given pivot point?

6. How many unknown forces can be solved for by applying a free-body diagram to a rigid body that is in equilibrium in one plane?

7. Describe the three rules for drawing a free-body diagram.

Calculation exercises

Exercise 4.a

Two light rods of equal length, **AB** and **BC**, are hinged at **B**. At their free ends they have frictionless rollers resting on a level floor. A light cord is tied from **A** to **C**.

Assume the assembly is stable in the plane of the paper. When a weight of 100 N is suspended from point **B**:

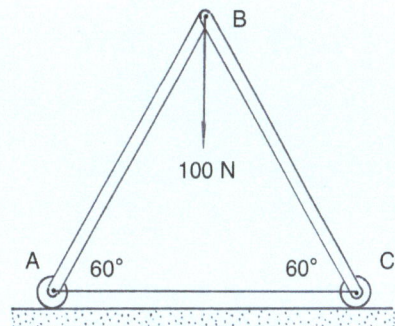

- What is the magnitude of the compression in each rod? [57.74 N]
- What will be the tension in the cord **AC**? [28.87 N]
- What additional assumption did you have to make in order to arrive at the above answers? [that the cord does not stretch]

Exercise 4.b

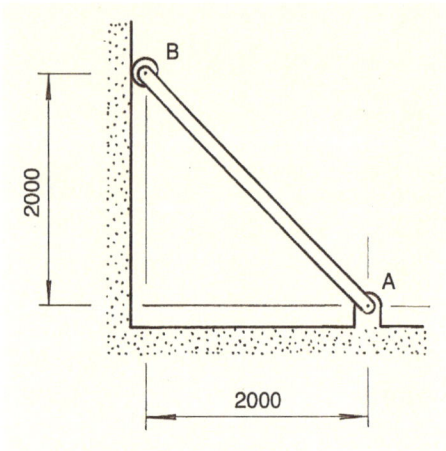

A 100 kg steel rod is hinged at end **A**, and has a light roller with negligible friction pivoting at end **B**, where the roller rests against a wall.

Draw a free-body diagram of the rod, showing all the forces acting on it.

Determine the magnitude and direction of the force that the hinge pin at point **A** exerts on the rod. [1097 N, W 63.4° N]

Exercise 4.c

A two metre long wooden pole, of density 800 kg/m³ and diameter 160 mm, rests on a level floor.

One end of the pole is raised by a cord that passes over a sheave and supports a mass-piece of weight W.

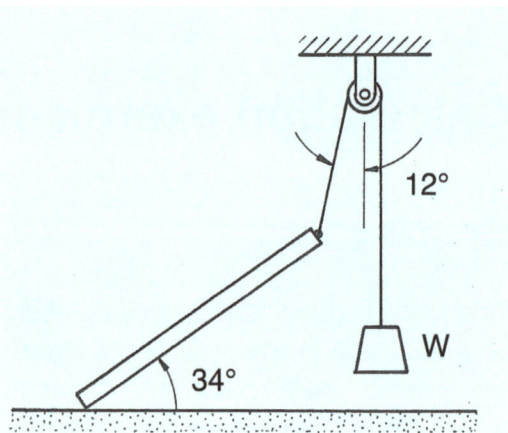

The pole reaches equilibrium in the position shown. Assume the friction in the sheave is neglible.

Determine the weight of the pole [315.6 N], and the mass of the hanging mass-piece. [19.19 kg]

Exercise 4.d

A solid triangular steel plate, mass 20 kg, is hinged to a wall at **A**, has a roller mounted at **B**, and carries a point load of 120 N at **C**.

Determine the magnitude and direction of the wall's reaction at point **A**.

[392 N, W 53.76° N]

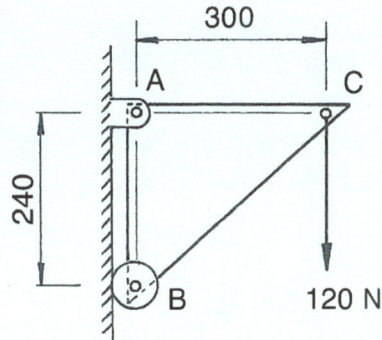

Exercise 4.e

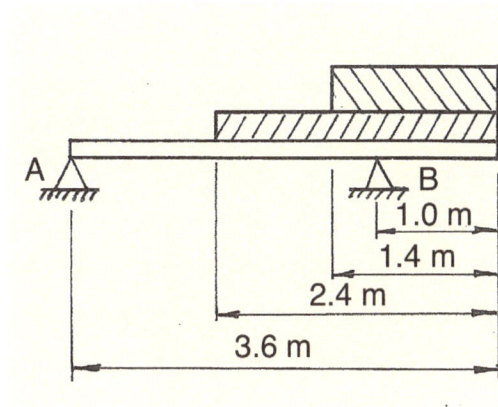

A beam rests on two simple supports, at **A** and **B**. The beam weighs 100 kN/m, and carries two distributed loads, a longer one of 50 kN/m and a shorter one of 80 kN/m.

Determine the magnitudes of the reactions at the supports.

[A = 107.1 kN; B = 484.9 kN]

Exercise 4.f

An L-shaped beam, whose weight can be ignored, is hinged at point **B** and rests on a roller support at point **A**.

A wire rope, attached at point **C**, passes over a frictionless sheave and carries a weight **W**.

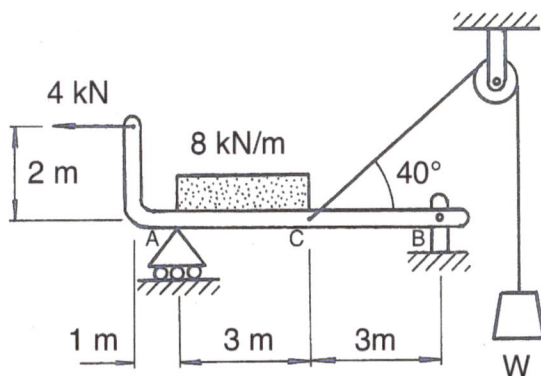

- Draw a free-body diagram of the beam, showing all forces resolved into vertical and horizontal components.
- If weight W = 20 kN, what will be the magnitude of the reaction at support **A**? [12.91 kN]
- Determine the value of weight W that will cause the reaction at **A** to be zero. [60.15 kN]

Exercise 4.g

A pole weighing 3 kN, hinged at its base, is in equilibrium under the action of the forces in the two ropes shown.

Assume there is minimal friction in the sheave.

Draw a free-body diagram of the pole, showing all forces resolved into vertical and horizontal components.

Determine the tension in the lower rope [3889 N] and the magnitude and direction of the reaction at the hinge. [6165 N, E 75.9° N]

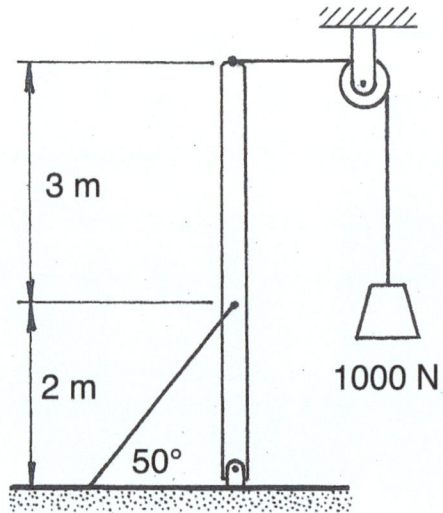

3 m

2 m

50°

1000 N

5

Centres of mass, centres of gravity and centroids

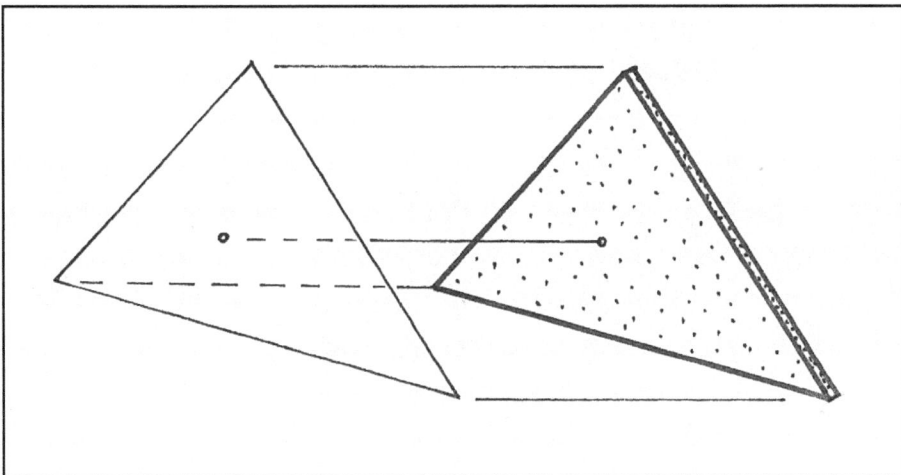

True/false test on this topic

	True/false Test # 5a Centres of mass, centres of gravity and centroids			
1	It is possible for the centre of gravity of an object to lie outside the material from which the object is made.	T	F	?
2	The centre of gravity of an object can be situated in a different position from that of its centre of mass.	T	F	?
3	Any calculations to determine the location of the centre of mass of an object are based on a procedure for determining the location of its centre of gravity.	T	F	?
4	The location of the centre of gravity of a rigid body does not change, unless the body is deformed or has material added or removed.	T	F	?
5	When determining the location of the centre of gravity of a complex-shaped flat plate of variable density, plate areas can be used in place of masses.	T	F	?
6	If some material is removed from an object, it is impossible to determine how the object's centre of gravity has shifted.	T	F	?
7	The centroid of a volume could lie at a different position than the centre of gravity of a solid object with the same dimensions.	T	F	?
8	To locate the centre of gravity of a structure made of straight rods, such as a truss, one first needs to know the location of the centres of gravity of the constituent rods.	T	F	?
9	It is not possible to determine the location of the centre of gravity of irregularly-shaped objects.	T	F	?
10	You can determine the mass of a long object such as a heavy beam, if you have only one weighing scale that can weigh up to just over half the weight of the beam.	T	F	?

Questions requiring short descriptive answers

1. Describe three circumstances in engineering practice in which it is important to be able to determine the location of a centre of gravity of an object.
2. Is the centre of gravity of an object always in the same position as the object's centre of mass? Explain.
3. Would the centre of gravity of an object always be located within the material of which the object is constructed? Explain.
4. Describe the relation between a centre of mass and a centroid.
5. Give an example of an object whose centre of mass does not coincide with its centroid.
6. Using one weighing scale with a maximum rating of 250 kg, how could you weigh a log whose weight is approximately 400 kg?
7. Explain why the loss of a gear tooth from a gearwheel that rotates at high speed could result in a vibration of the gearwheel.
8. Is the centre of mass of an object always in the same position as the centroid of the space occupied by that object? Explain.

Calculation exercises

Exercise 5.a

A steel rod, of length
$L = 2000$ mm and mass
30 kg, has two steel
collars fixed to it in the
positions shown.

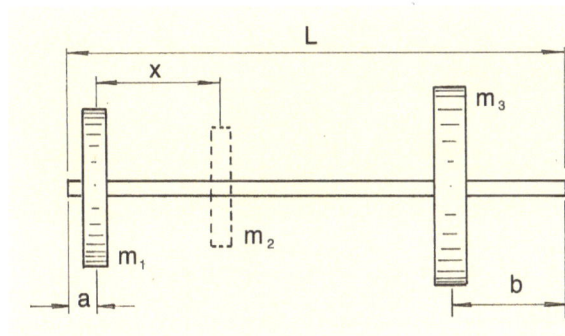

Their respective masses
are:
$m_1 = 20$ kg and $m_3 = 40$ kg.

Dimensions a = 100 mm and b = 400 mm

At what distance **x** should a third collar of mass $m_2 = 10$ kg be fixed so that the centre of gravity of the whole assembly lies at the midpoint of the rod? [0.4 m]

Exercise 5.b

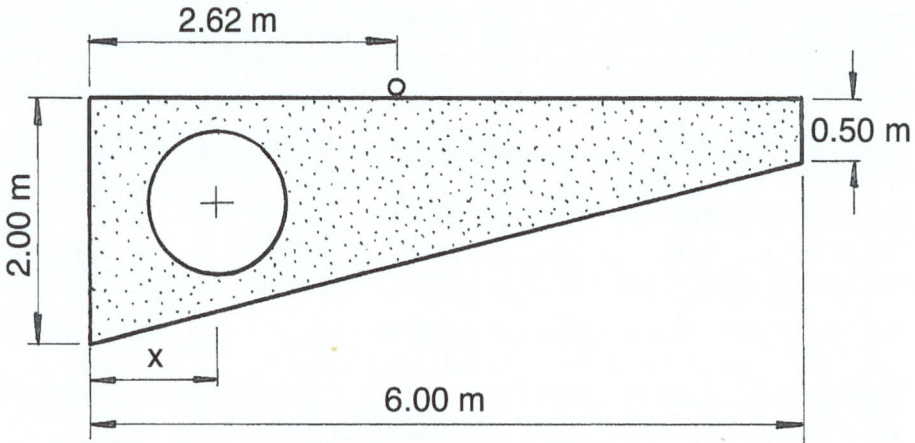

A reinforced concrete beam of uniform thickness and density has a lifting lug cast into it, in the position shown. When suspended from this lug, the beam has to hang with its upper edge horizontal.

In the process of casting this beam, a hole of diameter 1.20 m must be made in it, to accommodate a storm-water pipe which will pass through the beam after it is placed in position.

Determine the location of the centre of this hole with respect to the left hand edge of the beam, namely dimension 'x'. [1.161 m]

Exercise 5.c

A two-dimensional frame is made of steel bars welded together at their ends.

The table shows the weight per running metre of the materials from which each respective bar is made.

Determine:

- The lengths of all the bars, to the nearest mm, and their weights.
- The location of the centre of gravity of this frame relative to the x- and y-axis, to the nearest mm.

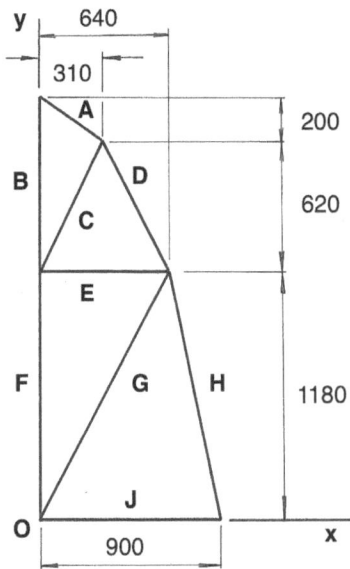

[X = 346 mm; Y = 801 mm]

Bar	length [mm]	Weight of material [N/m]	Weight of bar [N]
A		30	
B	820	30	
C		30	
D		30	
E	640	40	
F	1180	40	
G		40	
H		50	
J	900	50	

Exercise 5.d

A flywheel of total mass 60 kg consists of a rim, a hub and eight spokes. If one of the spokes were removed, and its mass was found to be 600 grams, how far off-centre has the centre of gravity of the wheel shifted? [by 1.485 mm]

Exercise 5.e

In this schematic diagram of an early race car, the large dots represent the major mass-concentrations that make up the vehicle. Determine:

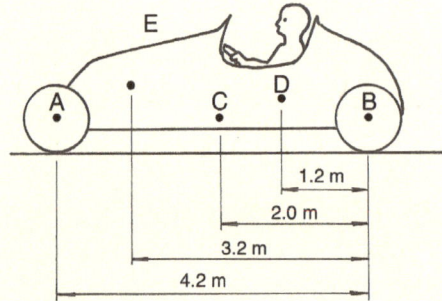

- The horizontal distance from the front axle to the centre of gravity of the car. [1.725 m], and

- The percentage of the total weight that is carried by the front wheels. [58.93%], given the masses:

    A: front wheel assembly and steering mechanism .... 100 kg
    B: rear wheel assembly .... 60 kg
    C: chassis, including differential .... 300 kg
    D: driver .... 80 kg
    E: engine .... 260 kg

6

Forces in structures

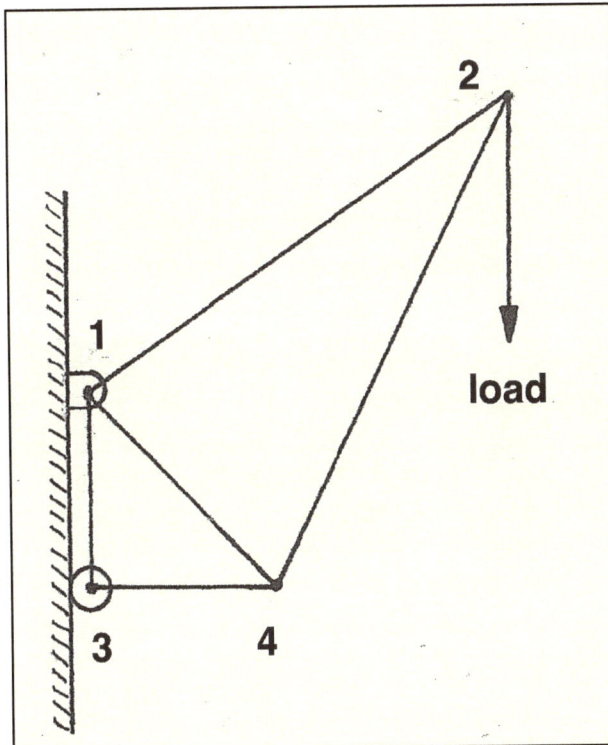

True/false tests on this topic

	True/false Test # 6a Forces in structures			
1	There is no difference between a truss and a frame, they are just different names for the same thing.	T	F	?
2	The force exerted by a member in a pin-jointed truss has to be orientated in the same direction as the member.	T	F	?
3	A roof 'truss' whose members are joined by nail plates is strictly speaking a frame, not a truss.	T	F	?
4	An over-constrained structure is one that has more than four unknown forces acting on it when under load.	T	F	?
5	The members of a pin-jointed truss are known as web elements.	T	F	?
6	For a member in a truss, timber is best used as a tie, not as a strut.	T	F	?
7	A truss whose outline consists entirely of quadrilateral spaces would be unstable.	T	F	?
8	Steel truss members are prone to buckling when used in compression.	T	F	?
9	All members of pin-jointed trusses should be designed to resist bending moments.	T	F	?
10	It is often possible to determine by inspection (i.e. without calculations) whether a given member will be a strut or a tie, under a given loading.	T	F	?

	True/false Test # 6b Forces in structures	T	F	?
1	There are three methods for determining the forces in the members of a pin-jointed truss under load.	T	F	?
2	A Maxwell diagram is a graphical representation of a truss.	T	F	?
3	A member of a pin-jointed truss that joins up in a 'T-junction' with two others that lie in a straight line, is probably not carrying a force.	T	F	?
4	A no-load member is always superfluous and has no purpose.	T	F	?
5	To determine the forces in the members of a loaded truss, it is essential to start by determining the values and directions of all the external forces acting on the truss under the given loading.	T	F	?
6	A well-constructed pin-jointed truss would not distort when under load.	T	F	?
7	If a 3-D truss is to be analysed in a similar way to a 2-D truss, all rigid members should be joined by ball and socket joints instead of pin joints.	T	F	?
8	A truss member that is always in tension could be replaced by a chain without changing the characteristics of the truss.	T	F	?
9	If a pin-jointed bridge truss has a node **A** that is vertically above another node **B**, then a vertical load **F** would have the same effect on the loading of all the other members whether it was applied at **A** or at **B**.	T	F	?
10	On a Maxwell diagram, the directions of all the forces are indicated by inserting arrow-heads on the force lines.	T	F	?

Questions requiring short descriptive answers

1. Describe the difference between a truss and a frame, using a diagram.
2. Define a 'pin-jointed' truss.
3. Why must the force carried by a member of a pin-jointed truss lie in the same direction as that member?
4. Describe the difference between a tie and a strut, in the context of load-bearing trusses.
5. Name and describe briefly the three methods of determining the values of the forces in the members of a pin-jointed truss.
6. What is a Maxwell diagram?
7. Can a 'no-load' member of a truss be identified by inspection? Explain with the use of a sketch.
8. If a member in a truss carries no load when a given external loading is applied to the truss, why is that member included?
9. Under what circumstances is it possible to determine the forces in the members of a three-dimensional truss? Explain.
10. For stability, trusses are usually made up of triangular spaces between the members. Is it essential for all the spaces within a truss to be triangular? Explain.

Calculation exercises

Exercise 6.a

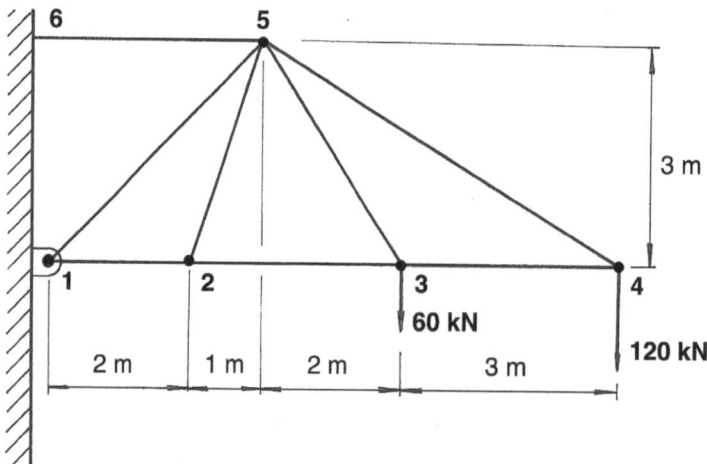

A pin-jointed truss is supported by a hinge at point **1** and a wire rope **5 - 6**.
It carries two loads, as shown. Ignoring the mass of the truss members, determine:

- The tension in the wire rope to keep the truss in equilibrium. [420 kN]

- The magnitude and direction of the reaction of the hinge pin at node **1**. [456.9 kN, E 23.2° N]

- By inspection, state which member is a no-load member, and explain why. [member **2 - 5**, as members **1 - 2** and **2 - 3** are in the same horizontal straight line, so would not carry forces with a vertical component]

- Does this member have a function, if it carries no load? [yes, if it was not there, there would be one long member **1 - 3** which would be in compression and might therefore buckle.]

- Draw a Maxwell diagram, using Bow's notation to determine the forces in members **3 - 4** and **4 - 5**. [233.3 kN in compression; 200 kN in tension]

Exercise 6.b

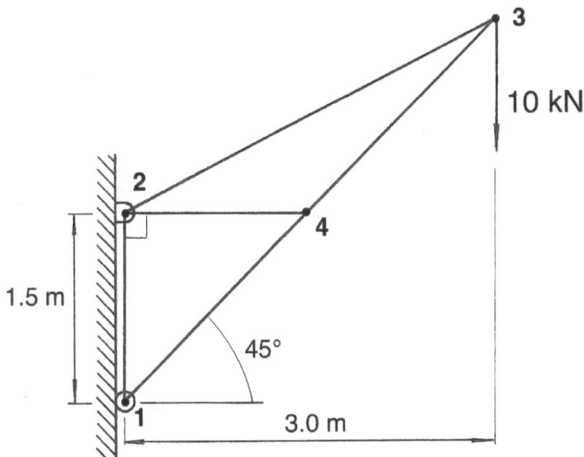

A simple wall crane consists of four members, pinjointed at nodes 1, 2, 3 and 4.

Ignoring the weight of the members themselves, determine the force in each respective member due to the given loading.

Assume there is no deformation in the structure when under load.

If your solution makes use of a Maxwell diagram with Bow's notation, the diagram should look as follows:

From the diagram, to scale:
forces in the respective members are:

1 - 2: 20.00 kN tension

2 - 3: 22.36 kN tension

3 - 4: 28.28 kN compression

4 - 1: 28.28 kN compression

2 - 4: zero

Since member **2 - 4** carries no load, is there a good reason for its inclusion? Explain.

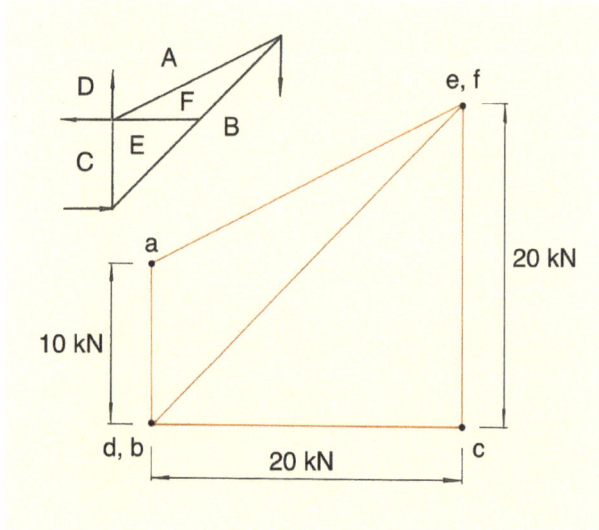

Exercise 6.c

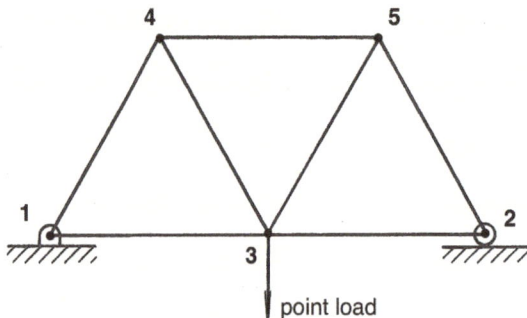

The diagram shows one side of a small pin-jointed bridge in the form of a Warren truss. The other side of the bridge is identical and parallel to this one. Assume that cross-bracing keeps the two sides stable in a vertical plane.

All the bridge members are 4 m long, and the bridge weighs 14 kN. A point load of 26 kN is suspended from node 3.

Analyse one side of the bridge only, to determine the tensions in each of the members shown.

[members **1- 3** and **3 - 2** are in tension: 5.774 kN;
members **1 - 4** and **2 - 5** are in compression: 11.55 kN;
members **3 - 4** and **3 - 5** are in tension: 11.55 kN;
member **4 - 5** is in compression: 11.55kN]

Exercise 6.d

A wall crane consists of four members, whose weight can be ignored, relative to the loading shown.

Determine which member carries the greatest load, and which one the smallest load, show the values of these loads, and state whether they are in tension or compression.

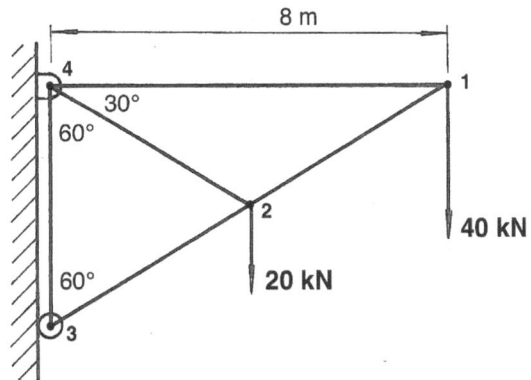

[The member carrying the greatest load is that between nodes **2** and **3**, which is in compression: 100 kN

The one carrying the least load is that between nodes **2** and **4**, which is in tension: 20 kN]

Exercise 6.e

Pole **Od** is 3 m long and is kept in a vertical position by three guy-ropes, anchored on a horizontal concrete base.

The anchoring positions are shown on a grid that consists of squares that are 2 m x 2 m.

The base of the pole is fixed

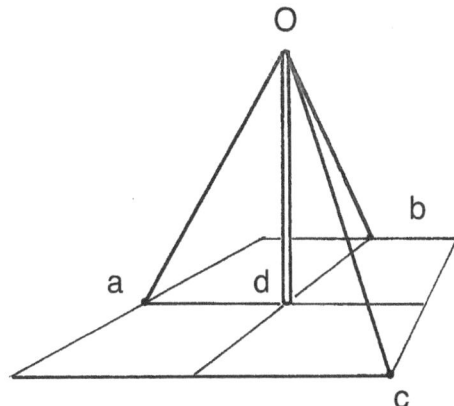

to a ball-and-socket joint that cannot transmit a force moment. The tension in rope **Oa** is 10 kN.

Determine the value of the tensions in the other two ropes and the compression in the pole.
[Tension in **Ob** is 10 kN; in **Oc** 11.44 kN; compression in pole is 24.96 kN]

7

Friction

True/false tests on this topic

	True/false Test # 7a Dry sliding friction			
1	Surface roughness is the only factor contributing to the amount of friction between two surfaces.	T	F	?
2	For two metals in contact, one contributory factor to the friction they experience could be molecular attraction.	T	F	?
3	The flexibility of one or both surfaces in contact affects the friction coefficient between them.	T	F	?
4	It is possible to determine the friction coefficient between two given materials to three decimal places.	T	F	?
5	The coefficient of static friction for two given surfaces is greater than the coefficient of kinetic friction for those surfaces.	T	F	?
6	When an attempt is made to slide one object relative to another, the friction force always opposes the attempted movement.	T	F	?
7	If an attempt is made to slide two surfaces in contact relative to one another, and there is no movement, the friction force between them must be zero.	T	F	?
8	The friction force generated between two surfaces is directly proportional to the amount of force pushing the surfaces together.	T	F	?
9	There are instances in which the friction coefficient has been found to exceed the value of 1.	T	F	?
10	The friction coefficient for aluminium on cast iron (unlubricated) is less than 0.3.	T	F	?

	True/false Test # 7b Dry sliding friction			
1	If you can stand on a rock that slopes at 35° without slipping, it means that the friction coefficient between your shoes and the rock must be greater than or equal to sin 35°.	T	F	?
2	The value of the limiting friction force can be determined only at the point of sliding.	T	F	?
3	A wide tyre on a car provides more friction than a narrow tyre of the same material.	T	F	?
4	If you have a crate resting on a slope and it needs a force of 50 N to push it down the slope, the limiting friction force must be 50 N.	T	F	?
5	In order to determine in theory, whether or not a block acted on by several forces will slide on a plane, it is necessary to make and test an assumption.	T	F	?
6	Friction loss is energy that can't be recovered after a mechanical movement has occurred.	T	F	?
7	The amount of 'friction loss' experienced is equivalent to the work done by the force that is attempting to cause the movement.	T	F	?
8	If the coefficient of friction between two surfaces was zero, then the angle of friction between those surfaces would be zero degrees.	T	F	?
9	If a plank bearing a brick is tilted gradually, and the brick begins to slide when the plank is at 40° to the vertical, the angle of repose is 50°.	T	F	?
10	Using the angle of friction simplifies the calculations in all types of dry (un-lubricated) friction situations.	T	F	?

Questions requiring short descriptive answers

1. Friction can be beneficial in some circumstances, and a hindrance in others. Describe briefly two examples of each.
2. Would you expect the friction coefficient between two given substances to be constant in all circumstances? Explain why, or why not.
3. Name three factors that contribute to the amount of friction between two plane surfaces in contact when an attempt is made to slide them relative to one another.
4. It is reported that the friction coefficient for a certain combination of metals can sometimes exceed the value of 1.0. If this is true, what could account for it to happen with metals and not between other substances?
5. What is the relationship between the angle of friction and the coefficient of friction? Describe.
6. In what circumstances is it more convenient to use the angle of friction than the coefficient of friction, to solve a friction exercise?
7. What is the angle of repose and how does it relate to the coefficient of friction?
8. Describe a way of determining the angle of repose for a brick on a wooden plank.
9. Consider dry sand being poured into a heap: why would some types of sand form a heap with sides more steeply angled than others? Explain.

Calculation exercises

Exercise 7.a

A large crate of mass 1600 kg is unloaded from a truck, by sliding it down a ramp. The coefficient of friction between the crate and the ramp is 0.3.

Determine the force parallel to the ramp, required to push the crate down the ramp, if the ramp is at 10° to the horizontal. [1912 N]

If the ramp is at 20° to the horizontal, determine the force that must be applied parallel to the ground, to prevent the crate from sliding down the ramp. [905 N]

Exercise 7.b

Two 10 kg steel plates are attached to rings, to each of which a cord is attached. They rest on a steel table, under a 50 kg mass-piece. The coefficient of friction is 0.5 for all surfaces. Determine:

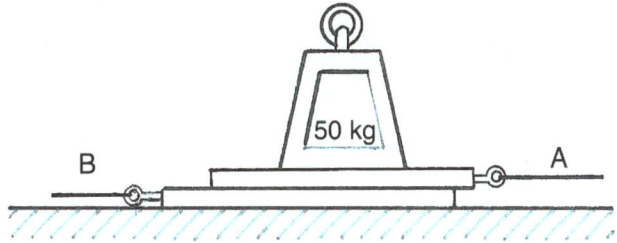

- The magnitude of the tension needed in cord **A** to slide the upper plate to the right, if cord **B** is restrained. [294.3 N]
- The magnitude of the tension needed in cord **B** to slide the lower plate to the left, if cord **A** is restrained. [637.7 N]

Exercise 7.c

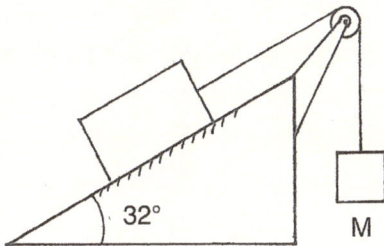

A block with mass 25 kg rests on an inclined plane. The block is connected to another mass, **M**, by a light cord passing over a frictionless sheave.

The coefficient of friction between the block and the plane is 0.45.

Determine the value of **M** required to just prevent the block from sliding down the plane. [3.708 kg]

Exercise 7.d

A block weighing 180 N rests on a horizontal surface, where the coefficient of friction between the block and the plane is 0.2.

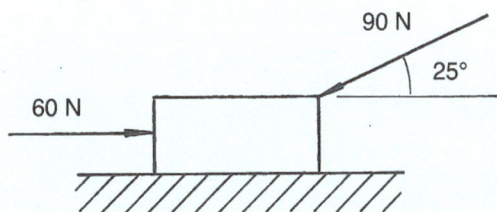

Determine by calculation whether or not the block will slide under the action of the two forces shown. [it does not slide]

Exercise 7.e

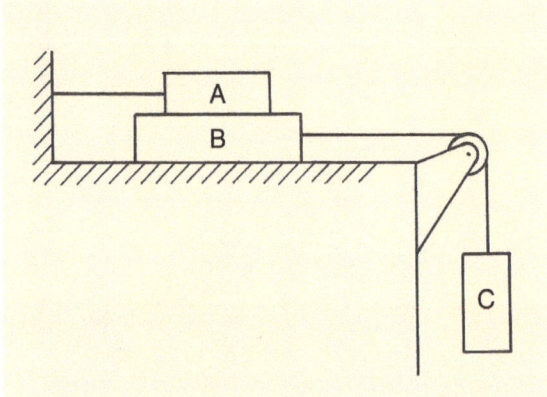

Blocks **A** and **B** are made of steel, with flat machined faces. Block **A** (12 kg) is connected to the wall by a light inextensible cord. Block **B** (20 kg) is connected to block **C** (also 20 kg) by a similar cord, passing over a sheave that is free to turn with negligible friction.

The coefficient of friction between block **B** and the table is 0.5. Determine the minimum value of the coefficient of friction between blocks **A** and **B** to prevent block **C** from moving downwards. [0.33]

If the value of the coefficient of kinetic friction between blocks **A** and **B** were only 0.1, what would be the downward acceleration of block **C**? [0.687 m/s^2]

8

Buoyancy

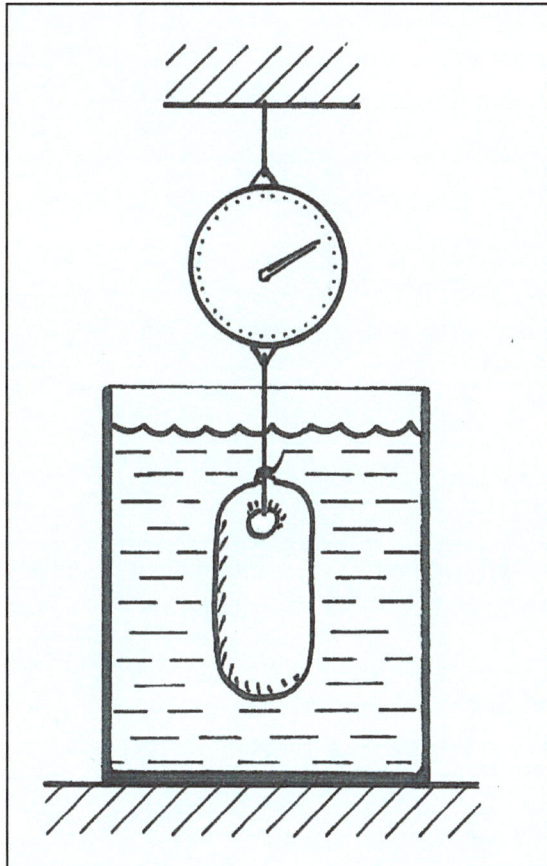

True/false tests on this topic

	True/false Test # 8a Buoyancy			
1	The magnitude of a buoyancy force can be calculated only when objects are completely immersed in a fluid, not when they are floating.	T	F	?
2	Hot-air balloons operate on a different principle to that of buoyancy.	T	F	?
3	When raising a bronze cannon from a wreck, when the cannon emerges from the water, the force in the rope suspending it will increase.	T	F	?
4	Buoyancy is the reason why divers have to use weight belts.	T	F	?
5	The density of a human being is very similar to the density of water.	T	F	?
6	Ships will not float in water, if the density of the material from which they are made is greater than that of water.	T	F	?
7	A gold coin would float in a bowl containing mercury.	T	F	?
8	The densest type of wood in the world (at 1200 kg/m^3) would float in the Dead Sea.	T	F	?
9	Neutral buoyancy means no tendency to move either up or down within the fluid.	T	F	?
10	A floating object displaces an amount of fluid whose weight is equal to its own weight.	T	F	?

	True/false Test # 8b Buoyancy	T	F	?
1	A closed-compartment object such as a sealed glass bottle containing air, will float, provided that its overall density is less than that of the fluid in which it is placed.	T	F	?
2	A beach ball pumped to a given pressure in air at a given temperature, when placed on cold water, will float higher than an identical ball placed on warm water.	T	F	?
3	When an open-topped vessel (e.g. roasting pan or rowing boat) floats, its total mass is equal to the mass of the water it displaces.	T	F	?
4	If a standard soccer ball has an overall volume of 5.792 litres and mass of 440 g, the downward force required to fully submerge it in fresh water will be less than 50 N.	T	F	?
5	To determine the lift provided by each cubic metre of gas in a hydrogen-filled balloon, it is necessary only to know the density of hydrogen and that of air, at the same temperature and pressure.	T	F	?
6	If an object suspended from a rope is denser than the fluid in which it is immersed, the tension in the rope suspending it is diminished by the same percentage, irrespective of the density of the object.	T	F	?
7	If a ship displaces 100 tonnes of sea water, that means the ship must have a mass of 100 tonnes.	T	F	?
8	A ship rated to have a displacement of 100 tonnes in sea water would also displace 100 tonnes of fresh water.	T	F	?
9	A ship on the sea is loaded until the water level is at a mark made on the hull. If this ship sails into fresh water, the water level will be below this mark.	T	F	?
10	An aircraft that is airtight will experience a buoyancy force due to the fact that it displaces its own volume of air.	T	F	?

Questions requiring short descriptive answers

1. Explain how to determine the overall density of a closed vessel such as a sealed, filled oil drum.
2. Explain the essence of Archimedes' principle.
3. How could one determine the density of an irregularly shaped object, whose volume cannot be deduced from measurements.
4. Describe how the position of the centroid of the submerged part of an open vessel such as a rowing boat, affects its stability.
5. State the requirement for a gas-filled balloon to achieve neutral buoyancy.
6. Explain why a ship made of steel, whose density is much greater than that of sea water, can float.
7. Describe the reasoning needed to determine the required density of a piece of wood that would float with two thirds of its volume protruding from the water.

Calculation exercises

Exercise 8.a

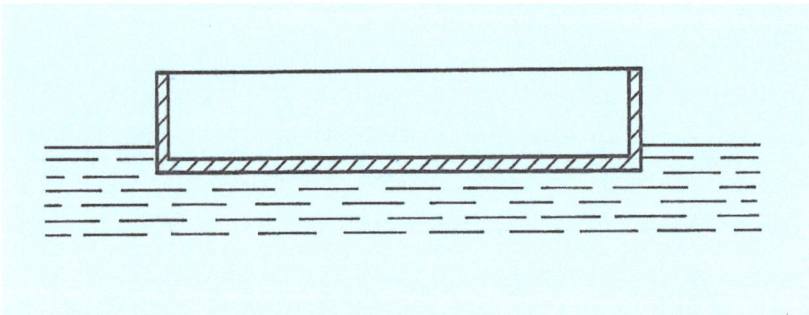

A rectangular plastic tray has outer dimensions 400 x 400 x 80 mm. When placed on fresh water, it floats with 10 mm of it submerged.

* What is the mass of the tray? [1.6 kg]
* If an object of mass 5 kg is placed in the floating tray, will it continue to stay afloat, or will it sink? [floats with 41.25 mm submerged]
* What is the maximum amount of mass that can be placed in the tray without causing it to sink? [11.2 kg]

Exercise 8.b

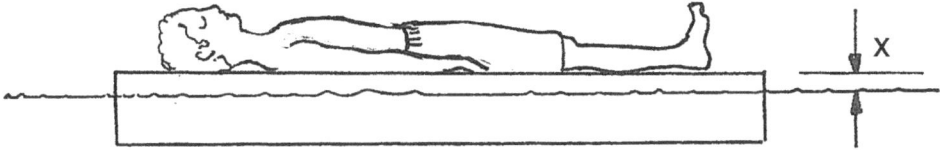

A boy of mass 50 kg lies down on a large polystyrene block, of density 25 kg/m^3, floating in a swimming pool. This block has dimensions 1500 x 500 x 200 mm.

- Determine the dimension **x**, the amount by which the block will protrude from the surface of the water when it is stable. [128.3 mm]
- Would this dimension be different if the same boy was floating on the same block on water on a planet where there was a different value for g, the gravitational acceleration? Explain. [no]

Exercise 8.c

A copper bar is fixed to one end of a wooden plank. When the assembly is placed in water, it floats.

This plank has dimensions 1000 x 250 x 20 mm and density 450 kg/m^3.
The copper bar is 250 x 60 x 20 mm and its density is 8900 kg/m^3.

Determine dimension 'y', the amount by which the plank sticks out of the water, if it is in fresh water [76 mm] and sea-water [102 mm].

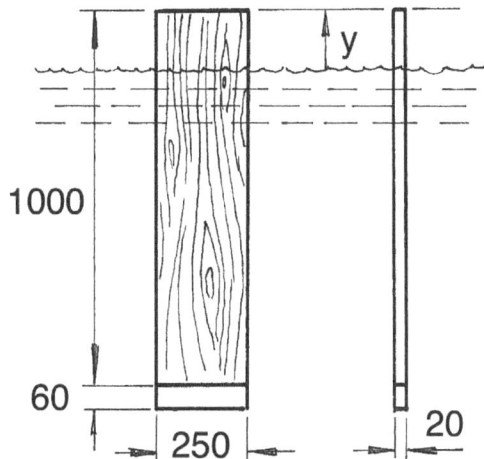

Exercise 8.d

A crane has to pull a concrete cube (density 2400 kg/m³) out of a freshwater lake. The cube sides have dimension **d**. While the cube is being raised, but still submerged, the tension in the crane rope: F = 275 kN.

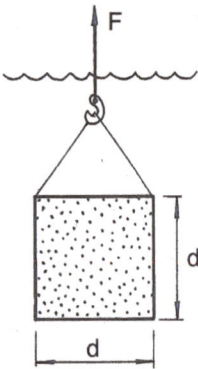

Determine:

• The value of d in metres. [2.715 m]

• The tension in the crane rope when the cube is out of the water. Take the density of air to be 1.29 kg/m³. [471.2 kN]

Exercise 8.e

A cylinder made of PVC (density 1390 kg/m³) is filled with expanded polystyrene (density 25 kg/m³) and sealed with two end-caps glued into place. The cylinder has outer diameter ϕ = 400 mm and wall thickness **t** = 3 mm. The overall length is **L** = 1200 mm and the thickness **d** of each end-cap is 20 mm. Determine:

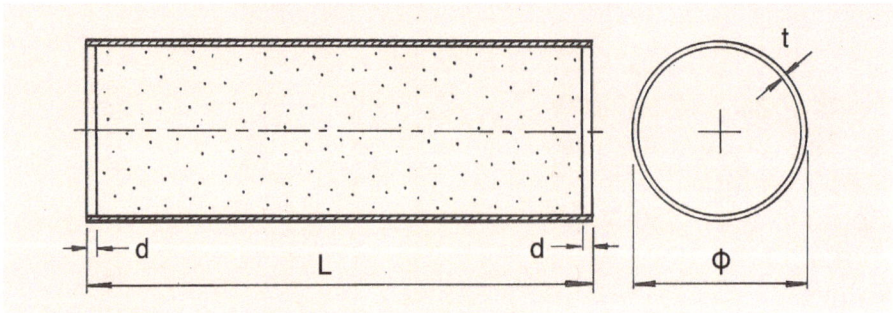

The mass of the cylinder. [6.241050 kg]
• The mass of the two end-caps. [6.778867 kg]
• The mass of the enclosed polystyrene. [3.535740 kg]
• The overall density of the assembly. [109.788110 kg/m³]
• The percentage of the volume of this assembly that will be submerged when it floats in sea water, of density 1027 kg/m³. [10.69%] and
• The amount of mass the assembly can support before it becomes submerged. [138.3 kg]

9

Linear motion with uniform acceleration

Impression of an early Buick race car

True/false tests on this topic

	True/false Test # 9a Linear motion with uniform acceleration			
1	If an object travels along a straight line from a given starting point, at no time could the distance travelled by the object be the same as its displacement from the starting point.	T	F	?
2	'Speed' is defined as the rate at which an object covers distance.	T	F	?
3	If an object's speed remains constant, but it is undergoing a change of direction, then the object must be accelerating.	T	F	?
4	On a velocity-time graph, when the velocity is in the positive direction, the graph has to have a positive slope.	T	F	?
5	On a velocity-time graph, the sum of all the areas between the graph and the time axis indicates the displacement from time t = 0.	T	F	?
6	A negative slope on a velocity-time graph indicates negative velocity.	T	F	?
7	The physical point on the line of motion, where the motion begins, can be found on a velocity-time graph.	T	F	?
8	On a graph of displacement vs time, the slope indicates the velocity.	T	F	?
9	If a particular linear motion can be described by a line that starts above the time axis on a velocity-time graph, then the object undergoing the motion must have started with a positive value of acceleration.	T	F	?
10	If an object travels in one direction on a straight line path, its displacement from the starting point at time t is given by (average velocity) × (time t).	T	F	?

	True/false Test # 9b			
1	The three equations of motion for straight line motion with uniform acceleration can only be applied for a period of motion whose initial and final conditions have been specified.	T	F	?
2	If the acceleration of an object has a negative value, it means the object is slowing down.	T	F	?
3	The velocity vs time graph for the motion of an electric motor vehicle accelerating from rest down a straight road starts off steeply, with the slope gradually diminishing.	T	F	?
4	Suppose two cars are travelling at the same speed. If car **A**'s brakes result in twice the deceleration that car **B**'s brakes can manage, then car **A** can come to a stop in one quarter of the distance that car **B** takes to stop.	T	F	?
5	The downward acceleration of a falling object is independent of its mass.	T	F	?
6	The acceleration of a moving object is equal to the slope of the velocity vs. time graph at that instant.	T	F	?
7	In a given exercise, any displacement to the right must always be allocated a positive sign.	T	F	?
8	It is impossible to be moving in the positive direction while experiencing a negative value of acceleration.	T	F	?
9	Suppose a small aircraft can accelerate up to its take-off speed of 108 km/h at a constant rate of 1.2 m/s^2. This aircraft could take off safely from a runway that is 400 m long.	T	F	?
10	If a vehicle's maximum acceleration is 2 m/s^2, and its maximum deceleration is 4 m/s^2, the shortest time in which it could cover a distance of 200 m in a straight line would be less than 20 seconds.	T	F	?

Questions requiring short descriptive answers

1. Describe the difference between 'distance' and 'displacement'.
2. Is 'speed' a scalar or a vector? Explain.
3. Define 'acceleration'.
4. Sketch a velocity-time graph for the motion of an object that starts at rest and moves with constant acceleration in the negative direction for a given time 't'.
5. Sketch a velocity-time graph for an object moving with an initial steady velocity of +5 m/s, and after 10 seconds it begins to accelerate in the positive direction with steadily increasing acceleration, for a further 5 seconds.
6. What quantity can be deduced from the slope of a graph of displacement vs time? Show by means of a sketch.
7. If the area below a velocity-time graph is negative, what does that imply for the motion the graph describes? Explain.
8. Can one use the three equations of linear motion to analyse the motion of an object, if there are phases of motion which have different values of acceleration? Explain with the aid of a sketch.
9. A skydiver in free fall might sometimes reach terminal velocity. What does that mean?

Calculation exercises

Exercise 9.a

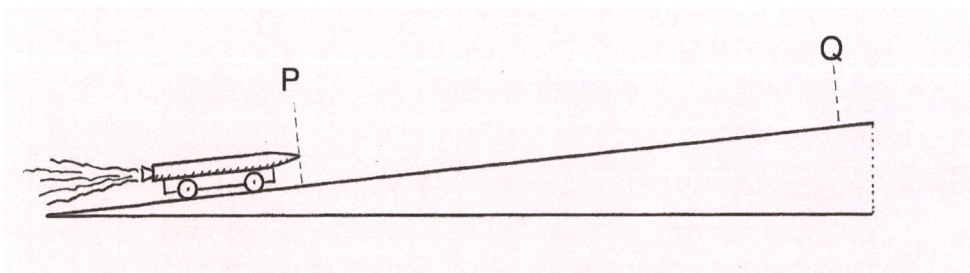

A rocket-powered trolley, on a straight rail going up a uniform slope, blasts off from rest at point P, accelerating at 8 m/s². After ten seconds, the rocket motor is switched off.

When not under power, the trolley experiences a downhill acceleration of 2 m/s^2.

Draw a velocity-time graph to represent the motion of the trolley form the time the rocket motor starts, to the time when the trolley passes through point P again, going downhill. Using the graph together with the equations of motion, deduce:

- The maximum velocity reached [80 m/s],
- The furthest distance from point P that the trolley reaches [2000 m],
- The velocity of the trolley when it passes point P on its way down [89.44 m/s downhill], and
- The total time taken, from starting at point P, to returning to point P [94.72 sec]

Exercise 9.b

Two cars are travelling in the same direction along a straight road, both doing a steady 90 km/h. Car **A** is 80 m behind car **B**.

At time t = 0, car **A** begins to accelerate, to overtake car **B**. Car **A** accelerates uniformly to a speed of 144 km/h over four seconds, and then slows down uniformly to 126 km/h over the next five seconds.

At time t = 2 seconds, the driver of car **B** notices that car **A** is attempting to overtake, and begins to decelerate at a uniform 1 m/s^2. Car **B** maintains this deceleration until t = 9 seconds.

Draw a velocity-time graph to scale, showing the motion of both vehicles from t = 0 until t = 9.
Determine the position of car **A** relative to car **B** at time t = 9. [**A** is 37 m ahead of **B**]

Exercise 9.c

This velocity-time graph illustrates the movement of two vehicles, **A** and **B**, which start from rest, from the same position on a straight, level road. Vehicle **A** departs at time t = 0, and vehicle **B** departs 15 seconds later.

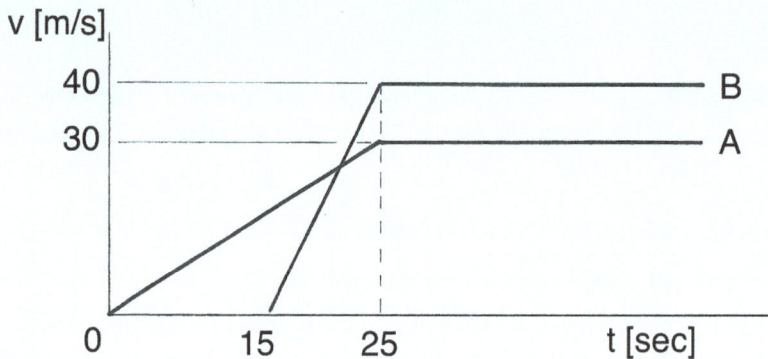

- At what time are the two vehicles level with one another? [t = 42.5 sec]
- How far down the road from the starting position does **B** draw level with **A**? [900 m]
- How many seconds after **B** overtakes **A**, will **B** be 400 m ahead of **A**? [40 sec]

Exercise 9.d

A train with mass 77.4 tonnes travels between two stations which are 14.1 km apart. It can accelerate uniformly from rest such that it will reach its top speed of 108 km/h in 3 minutes and 40 seconds. It can also decelerate uniformly while braking, from top speed to a dead stop in a distance of 2.7 km. Determine:

The shortest time in which this train could complete the journey [11 min and 10 seconds], and
The amount of work that is done against friction during the deceleration of the train [34.83 MJ]

Exercise 9.e

This graph of velocity vs time represents a short journey taken by a car on a straight, level road. It accelerates from rest, through three gears, continues at constant velocity for a while in 4th gear, then brakes with uniform deceleration until it comes to rest.

If it travelled for 360 m in 4th gear, what was its velocity then? [108 km/h]
How far did it travel from start to stop? [963 m]
What was its average velocity for the entire trip? [72.23 km/h]

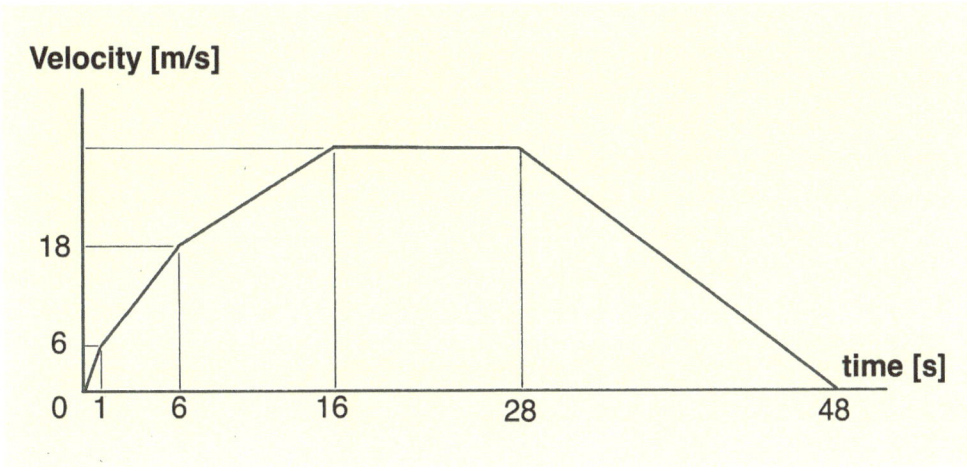

Exercise 9.f

Two trains on parallel tracks travel from start to stop, from station **1** to station **2**, a distance of 18 km.

Train **A** leaves at 06:00, accelerates at 0.5 m/s^2 until it reaches 90 km/h, then keeps up this speed until it approaches station **2**, when it decelerates uniformly to a stop over the last 1200 m of the journey.

Train **B** leaves at 06:03, accelerates at 0.6 m/s^2 until it reaches 144 km/h, after which it maintains this speed. When it approaches station **2**, it decelerates uniformly at 0.4 m/s^2, to stop exactly at the station.

Draw to scale a velocity-time graph for the motion of these two trains (for 12 minutes altogether) and from the graph determine:

*   The time taken by train **A** to complete the journey. [10 min 55 sec]
*   The time of day when train **B** overtakes train **A** [06:08:25], and
*   How far they are from station **1** when **B** overtakes **A**. [12 km exactly]

59

10

Motion influenced by gravity

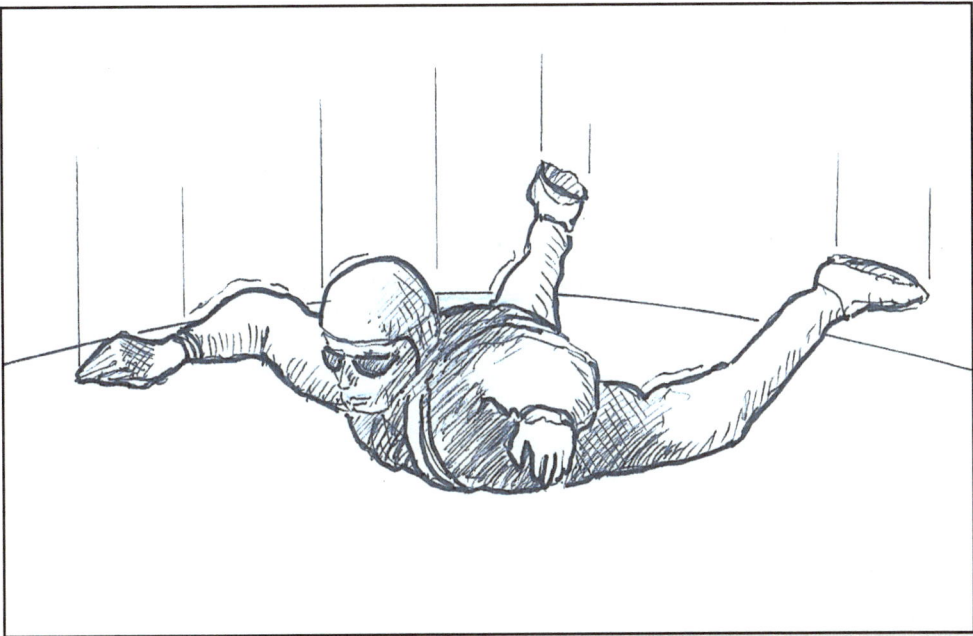

True/false test on this topic

	True/false Test # 10a Motion influenced by gravity			
1	The reason that a feather dropped from a given height takes longer to reach the ground than does a small stone with the same weight as the feather, is that the respective air resistances on the two objects differ.	T	F	?
2	The value of the gravitational acceleration, g, is always 9.81 m/s^2, at all locations on Earth.	T	F	?
3	For a falling object, the terminal velocity is that velocity at which the air resistance to its motion is equal to the weight of the object.	T	F	?
4	The height of a cliff can be estimated by dropping a stone from the top of the cliff and measuring the time it takes to reach the ground below.	T	F	?
5	The trajectory of a projectile can be determined if the launch speed is known.	T	F	?
6	Air resistance plays a minimal role in affecting the trajectory of a high speed projectile.	T	F	?
7	Ignoring air resistance, a projectile fired at a given speed over horizontal ground will have the same range when the launch angle is 23° as when it is 67°.	T	F	?
8	The effect of air resistance on a projectile is always more noticeable in the vertical direction than in the horizontal direction.	T	F	?
9	It is impossible to determine the launch speed of a projectile such as an arrow, by purely mechanical means.	T	F	?
10	If James and Peter both throw a cricket ball at the optimum trajectory, and James can achieve a throw distance of 80 m, while Peter can achieve only 60 m, then the difference in their launch speeds must be 20 m/s.	T	F	?

Questions requiring short descriptive answers

1. Explain how you could estimate the height of a cliff by dropping a stone from the cliff-top to the ground below.
2. Explain why the trajectory of a projectile such as a golf ball is not a perfect parabola.
3. If you were shooting an arrow over flat ground to achieve the furthest achievable distance in range, would you try to launch it at 45°, or a bit above that angle, or a bit below that angle? Explain.
4. In what way does air resistance have a desirable effect on the trajectory of a golf ball? Explain.
5. The distance that a javelin can be thrown depends on two factors. Explain what they are, and how they affect the distance achieved.
6. Explain how the launch velocity of a crossbow bolt can be determined by a purely mechanical means.

Calculation exercises

Exercise 10.a

A firework rocket takes off vertically upwards, accelerating at 12 m/s². The gunpowder that propels the rocket burns out in 3 seconds. Ignoring the effect of air resistance, determine:

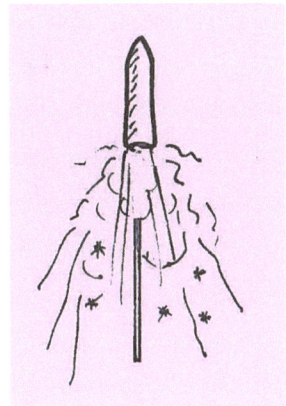

- The maximum velocity reached by the rocket [36 m/s],
- The maximum height reached [120.1 m] and
- The velocity with which the rocket hits the ground [48.53 m/s]

Knowing that air resistance very much affects the flight of objects, which of your three answers do you find to be the most unreliable? Explain why.

Exercise 10.b

On the surface of the moon, where $g = 1.6$ m/s^2, a small projectile is launched vertically, with an initial velocity of 40 m/s, from a platform that is 50 m above the ground. On its way down it falls to the moon's surface, narrowly missing the platform, but effectively keeping to a vertical path. Determine:

- The maximum height above the moon's surface that the projectile reaches. [550 m]
- The time taken from launch to landing. [51.22 seconds]
- The velocity of the projectile 36 seconds after being launched. [17.6 m/s downwards]
- If a small part of the projectile breaks off, 8 seconds after the launch, which of the two reaches the ground first, the projectile or the loose part? Explain your answer. [This answer is deliberately not provided, to promote discussion.]

Exercise 10.c

At time $t = 0$, a sack of grain is dropped from the top of a building 120 m high. Three seconds later, an arrow is shot at the falling bag from the same launch point, with an initial velocity of 60 m/s. Assuming the arrow is properly aimed, would it be possible for it to hit the sack before the sack reaches the ground?

[Yes, it could. They are level at $t = 4.444$ seconds, wheras the sack would hit the ground at $t = 4.946$ seconds]

Exercise 10.d

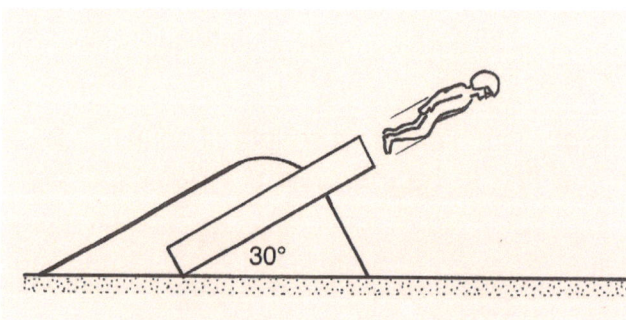

A 'human cannonball' of mass 70 kg is launched from the 'cannon' barrel by a plunger driven by compressed air.

The force that the plunger exerts on his feet varies from a maximum value to half that value over the 2.8 m of his travel while in the barrel.

The maximum upward force on his feet that he can endure is four times his weight. Ignore the effect of air resistance on his trajectory, and determine

- How much work is done on the man to accelerate him up the barrel [5768 J],
- His launch velocity, if all the work done on the man is converted to kinetic energy [12.84 m/s],
- The maximum height above the launch point that he reaches on his trajectory [2.100 m], and
- The range of this trajectory, namely, how far from the launch point he would again be at the same altitude as the launch point [14.55 m].

Exercise 10.e

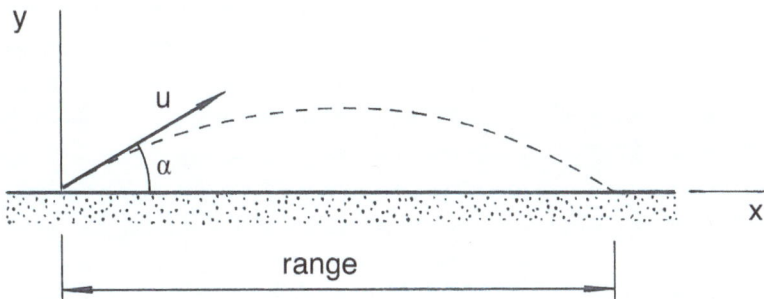

A cannon ball is fired with an intial velocity of u = 200 m/s, at an angle of elevation of α = 30° over flat ground. Ignoring the effects of air resistance, determine:

- The height of the ball above the ground 8 seconds after firing. [486.1 m]
- The angle at which the ball is moving at that instant. [7.083° upward of horizontal]
- The range of this cannon when fired at this elevation. [3531 m]

Exercise 10.f

Two arrows are fired vertically upwards at the same instant. Arrow **A** is fired from the ground, with an initial velocity of 60 m/s. Arrow **B** is fired from the top of an adjacent 20 m high building, with initial velocity **u**. Ignoring the effect of air resistance, determine:

- The value of **u** such that both arrows reach the top of their flight paths at the same height above the ground [56.64 m/s],
- Which of them reaches this height first, and by how many seconds? [**B** first, by 0.343 sec]

11

Rotational motion with uniform acceleration

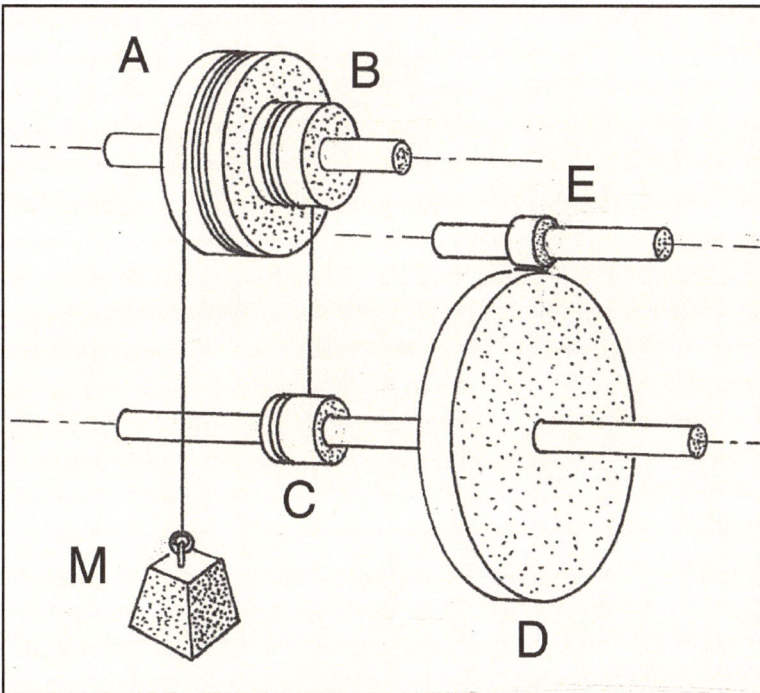

True/false tests on this topic

	True/false Test # 11a Rotary motion with uniform acceleration			
1	The radian, as a measure of angle, has been chosen arbitrarily.	T	F	?
2	There are more than six radians in one full revolution.	T	F	?
3	For the purposes of applying the equations that govern rotational motion, angular displacement is measured in degrees.	T	F	?
4	Angular velocity is a vector quantity.	T	F	?
5	If the angular acceleration of a rotating wheel is negative, it means the wheel is moving in the opposite direction to that in which it started moving.	T	F	?
6	If a treadle-powered grindstone of diameter 800 mm needs to reach a peripheral speed of 5 m/s in order to be able to sharpen a tool, and this takes 20 seconds from a standing start, the angular acceleration has been 0.625 rad/s^2.	T	F	?
7	Angular motion can be analysed by the use of a graph of angular velocity vs time.	T	F	?
8	In a gear drive, the angular velocities of two meshing gears are in direct proportion to their pitch circle diameters.	T	F	?
9	When two gears mesh, their gear ratio is inversely proportional to the numbers of teeth they respectively possess.	T	F	?
10	In a belt drive, the diameter of a tensioner (or idler) pulley makes no difference to the velocity ratio achieved between driver and driven pulley.	T	F	?

	True/false Test # 11b **Rotary motion with uniform acceleration**			
1	The velocity ratio of a drive is the ratio between the angular velocities of the input and output gears.	T	F	?
2	If a drive train consisting of two meshing gears is accelerated from rest, the ratio of the angular accelerations of the input and output gears is identical with their velocity ratio.	T	F	?
3	The linear speed of a point on the surface of the Earth, due to the Earth's rotation, depends on the longitude of that point.	T	F	?
4	If a gearbox has a gear ratio of 12, it follows that the angular velocity of the output shaft will be 12 times that of the input shaft.	T	F	?
5	The length of a given arc is equal to the angle it subtends in radians, divided by the radius.	T	F	?
6	The equations that govern the relations between angular velocity, angular acceleration, angular displacement and time have an identical format to those of their counterparts that govern linear motion.	T	F	?
7	Assuming a circular orbit, as an approximation, in order to determine the speed of the Earth in its orbit around the sun, it is necessary to know the distance from the Earth to the sun.	T	F	?
8	If a rotatable drum is accelerated up to a known angular velocity and then allowed to freewheel until it comes to rest, it is possible to determine how many revolutions it made while slowing down, by timing the period of deceleration.	T	F	?
9	A mass-piece is suspended from a cord wound around the horizontal axle of a circular disc that is able to rotate, but is restrained by a brake. If the brake is released to allow the mass-piece to descend, thus turning the axle: timing the descent over a known distance will enable one to determine the angular acceleration of the disc.	T	F	?
10	On a graph of angular velocity vs time, the area under the graph represents angular acceleration.	T	F	?

Questions requiring short descriptive answers

1. Define a radian and show why there are 2π radians in a circle.
2. List the similarities for what can be deduced from a velocity-time graph for linear motion and one for angular motion.
3. When two gearwheels mesh, we determine the ratio of their angular velocities by inverting either the ratio of their numbers of teeth, or the ratio of their pitch circle diameters. Explain why the latter two ratios are the same.
4. Describe the circumstances under which it is preferable to use a chain drive rather than a belt drive for transferring rotation from one shaft to another.
5. What is the function of an idler (or tensioner) pulley in a belt drive? Explain with the aid of a sketch.
6. Does the diameter of a tensioner pulley affect the belt speed? Why, or why not?
7. Show by means of a sketch why the ratio of the angular accelerations of two meshing gears is identical with the ratio of their angular velocities.
8. Describe the difference between a simple gear train and a compound gear train.

Calculation exercises

Exercise 11.a

A flywheel of diameter 0.7 m is accelerated uniformly from rest to a speed of 1000 rev/min over a period of 10 seconds. The power is then switched off, and the flywheel slows down gradually and uniformly to rest over a further 4 minutes. Determine:

- The total number of revolutions turned by the wheel in the 4 min and 10 seconds of motion. [2083]
- The angular velocity of the wheel (in rad/s) one minute after the start of the motion. [82.9 rad/s]
- The greatest tangential velocity reached by a point on the surface of this wheel at any time during this motion. [36.65 m/s]

Exercise 11.b

A cord attached to a weight **W** is wound around shaft **A**, which is fixed concentrically to a gearwheel **B**. This shaft is free to turn in bearings, (not shown). Gearwheel **C** on its own shaft **D** is also free to turn, and meshes with gearwheel **B**.

Weight **W** is released from rest, and descends a distance of 1.4 m in 1.68 seconds, after which the cord comes off the shaft **A**.

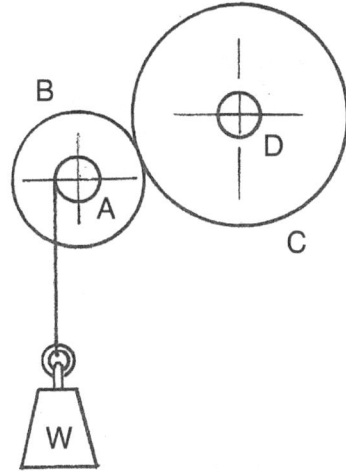

Diameter of shaft **A**: 20 mm. Pitch circle diameters: **B**: 240 mm, **C**: 360 mm

Determine:

- The linear acceleration of weight **W** before the string comes loose [0.9921 m/s^2],
- The angular acceleration of shaft **A** [99.21 rad/s^2],
- The linear acceleration of a point on the pitch circle of gear **B** [11.90 m/s^2], and
- The maximum linear velocity attained by a point on the rim of gearwheel **C** [20.00 m/s].

Exercise 11.c

A new theatre being built requires a revolving stage, of diameter 12 m. This large disc is to have various rotation patterns. One of them is specified as follows:

It must make one full revolution in sixty seconds, speeding up uniformly over five seconds and slowing down uniformly over five seconds before coming to a stop. Determine:

- The maximum linear speed reached by a point on the rim of the disc [0.6584 m/s], and

- The value of the angular acceleration required, in rev/min/sec. [0.218 rev/min/sec]

A different rotation pattern requires the edge of the rotating part of the stage not to exceed 0.3 m/s, while the acceleration and deceleration have to be 0.1 rad/s². How long will it take for the stage to turn one full revolution? [127.7 seconds]

Exercise 11.d

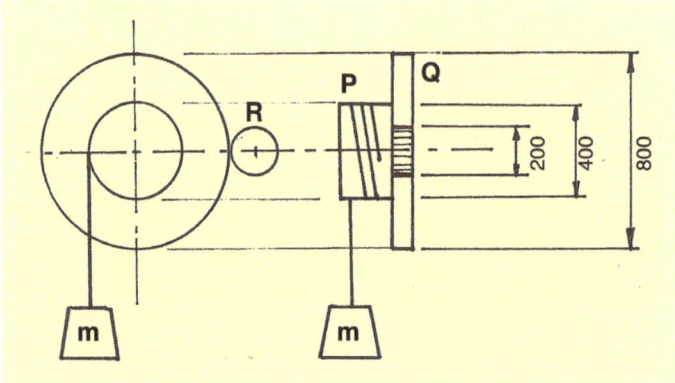

Drum **P** is fixed to a large gearwheel, **Q**, which meshes with a smaller gearwheel, **R**. Mass-piece **m** is attached to a cord wound around the drum. The length of cord on the drum is 8 metres.

When mass-piece **m** is released from rest, it descends the first 4.0 m in exactly 2 seconds. When the cord is fully unwound, it comes off the drum, allowing the mass-piece to fall, after which the drum slows down uniformly to rest, over a further 12 seconds.

Draw a graph of angular velocity vs. time for drum **P**, and determine:

- The maximum angular velocity of gearwheel **Q** at any time during the motion of the drum, [20 rad/s] and
- The number of revolutions turned by gear **R** from the start to the end of the motion. [89.13 revs]

Exercise 11.e

If the average radius of the Earth (if presumed spherical) is 6371 km, determine the shortest distance, measured along the surface, from the

Equator to a point with latitude 44° N, to the nearest km [4893 km].

Determine also the linear (tangential) speed of that point, due to the Earth's rotation, to the nearest km/h. [1200 km/h]

Exercise 11.f

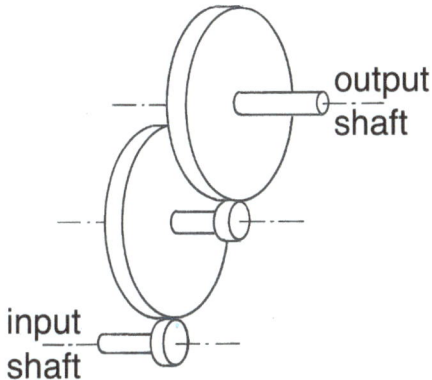

The velocity ratio of a compound gear train is 20. If the motor turning the input shaft accelerates from rest to its operating speed of 1500 r/min in 0.8 seconds, determine:

The angular acceleration of the output gear [9.817 rad/s^2], and

The number of revolutions made by the output gear in the first 2 seconds of motion. [2.0 revs]

Exercise 11.g

In this apparatus for raising or lowering mass-piece **M**, all shafts are mounted in bearings, (not shown) in a sturdy machine frame.

Drive shaft **E** carries a friction wheel that engages with the perimeter of disc **D**. Assume no slip.

Effective diameters [mm] are:

A: 400
B: 200
C: 150
D: 600
E: 120

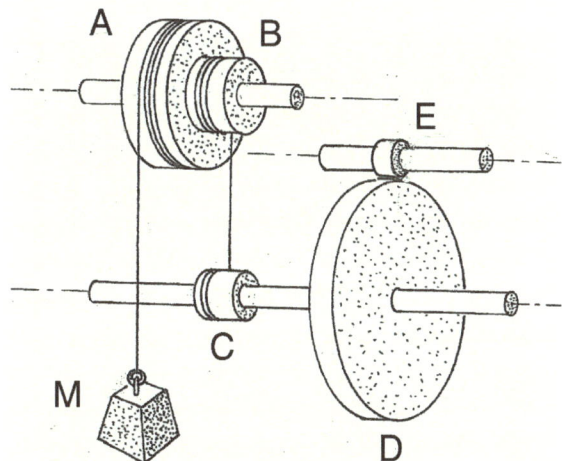

If mass-piece M needs to be raised at a velocity of 0.5 m/s, what should be the angular velocity of drive shaft **E**? [159.2 r/min]

12

Work, energy
and power

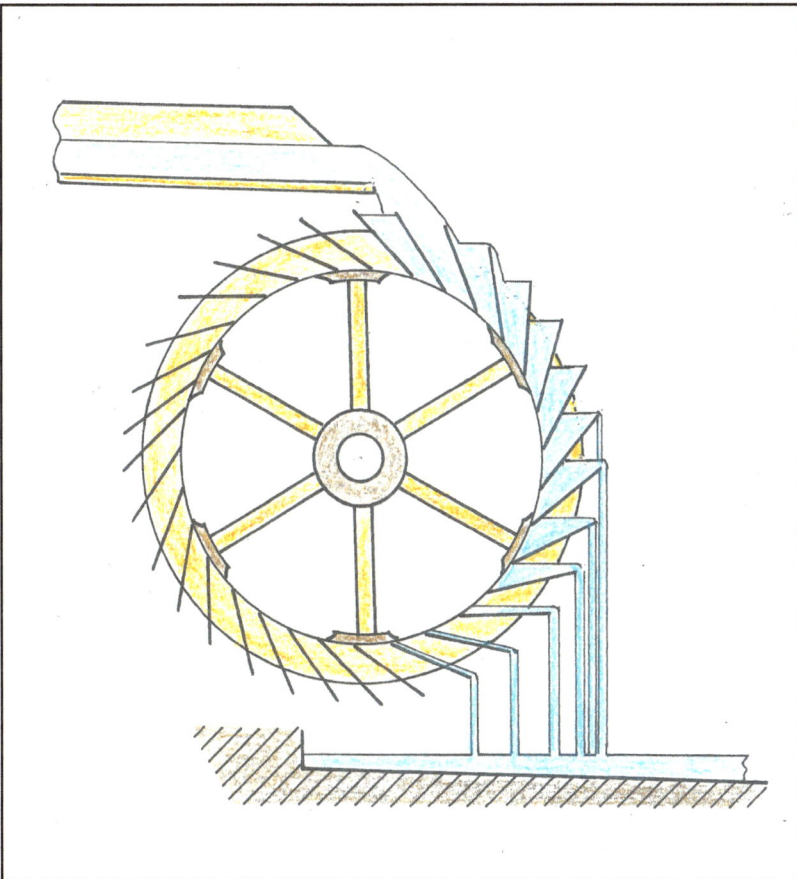

True/false tests on this topic

	True/false Test # 12a Work, Energy and Power	T	F	?
1	The Law of Conservation of Energy states that the gravitational potential energy of an object is always equal to its kinetic energy.	T	F	?
2	If you hold a heavy gym weight motionless above your head, you are performing work on the weight.	T	F	?
3	Work is the rate at which energy is expended.	T	F	?
4	The amount of mechanical work done by a constant force F, moving its point of application a distance 'd' in its own direction, is given by F(d).	T	F	?
5	If you exert a force P to push a force F back along its own line of action by a distance 's', the work done against force F is equal to F.s.	T	F	?
6	The mechanical energy stored in a spring is equal to (maximum force needed) × (distance that the spring is compressed).	T	F	?
7	If you weigh 700 N, and you run up a flight of stairs, raising your centre of gravity by 10 m, taking 7 seconds to do so, your power output has been 100 W.	T	F	?
8	The work done by a torque T that rotates an object through angle θ is equal to Tθ.	T	F	?
9	One of the outcomes of performing mechanical work on an object, could be to deform the object.	T	F	?
10	If a variable force acts on an object over a given distance, the work done by that force is represented by the area under the force vs distance graph of this action.	T	F	?

	True/false Test # 12b Work, Energy and Power			
1	When a bullet penetrates a large block of wood that is not observed to move as a result of the collision, all the kinetic energy of the bullet has been lost to heat.	T	F	?
2	To 'perform work against gravity' means exerting a force that is equal and opposite to the gravitational force that acts on an object.	T	F	?
3	When you attempt to push a heavy stone across a stone floor, before it starts moving you may be exerting a considerable force but performing no mechanical work.	T	F	?
4	When you raise a bucket of water from a deep well by means of a chain passing over an overhead sprocket, the rate at which you need to perform work steadily decreases while the bucket is being raised.	T	F	?
5	Suppose an archery bow is rated for a maximum pull of 180 N for a draw length of 700 mm: the energy stored in the bow at full draw would be more than 65 joules.	T	F	?
6	If you raise a weight of 100 N to a height of 16 m, the amount of work it could do by descending gradually while driving a mechanism that is 75% efficient, would be 1200 J.	T	F	?
7	Driving a turbine using the flow of water from a dam is an example of making use of gravitational potential energy.	T	F	?
8	The expression for the value of the kinetic energy of an object moving at velocity 'v' can be derived by considering the work done to accelerate the object to that velocity, from rest.	T	F	?
9	An airgun works by releasing the elastic energy stored in a volume of air that has been compressed.	T	F	?
10	A stream-shot waterwheel operates by making use of gravitational potential energy.	T	F	?

	True/false Test # 12c Work, Energy and Power			
1	The principle of Conservation of Energy can only be used to solve for unknowns in situations without impacts.	T	F	?
2	In any impact situation, the less elastic the colliding objects are, the less energy is lost as a result of the impact.	T	F	?
3	The unit of power called the horsepower was derived from research into how much power actual horses could expend.	T	F	?
4	A very fit adult could easily expend energy at a rate of 2 horsepower for about one minute.	T	F	?
5	Even if a vehicle engine is rated at (say) 200 kW, the engine will not be operating at this capacity unless the load demands of the situation require it.	T	F	?
6	The power transmitted by a known torque T will be constant at all speeds of rotation.	T	F	?
7	In order to describe the power output of a given person, you need to take into account the duration of time for which that person could sustain a specified level of energy output.	T	F	?
8	Consider a metal ball being either lowered gradually, or dropped onto a table: the reason that shock loading transmits more energy than gradual loading has to do with the total mechanical energy possessed by the ball before contact.	T	F	?
9	When you lower a load gradually, such a taking a package off a table and lowering it the floor, you are performing work.	T	F	?
10	If you raise a load using a chain passing over a single fixed overhead pulley, the mechanical work you need to do, per unit of distance that the load is raised, remains constant for the duration of the action.	T	F	?

Questions requiring short descriptive answers

1. Define 'mechanical work' in a sentence or two.
2. Explain why the amount of work done to compress a coil spring by a distance 'x' is proportional to x^2.
3. What is meant by doing work against a force?
4. Sketch a graph of force vs extension for the force exerted by a bowstring on an arrow, from the position where the arrow is just released from full draw, up to the position where the string is again taut.
5. Could an astronaut who is free-floating in space and not in touch with any other object, perform mechanical work? If so, how?
6. When the value of a force varies during the time for which it acts, how could we determine the amount of mechanical work it does?
7. Explain why an object that is raised acquires gravitational potential energy.
8. List three examples in which we make use of stored elastic energy.
9. Why is it not practicable to make use of the Law of Conservation of Energy when trying to analyse a situation in which an impact has occurred?
10. Explain the difference between energy and power.
11. Describe one method by means of which the power output of a human could be measured.
12. If a truck accelerates gradually while carrying a load up an incline, does it expend power at a constant rate? Why, or why not? Explain.
13. Explain why you are not performing mechanical work while you are standing still, while holding a 20 kg weight above your head.
14. When a water pumping system that pumps water from level A to level B is described as 45% efficient, what does that mean essentially?
15. A compression spring is rated to have a given value of stiffness, k, sometimes known as the spring constant. Would this value actually remain constant under all circumstances, assuming the spring is operating in the elastic range? Explain.
16. Name the variables that affect the amount of power that can be obtained from a waterwheel, and state what effect each of them has on the power output.

Calculation exercises

Exercise 12.a

A rowing machine in a gym requires the user to pull against the resistance offered by two parallel springs. It requires a force of 40 N to extend the two springs by 100 mm from their relaxed state.

If a gym user pulls hard enough to extend the springs by 400 mm, what is the maximum force she will need to exert at the end of the pull stroke? [160 N]

On the return stroke, she also has to perform mechanical work against the springs, in order to release their tension gradually. If she had to just let go, they would snap back dangerously.

If she carries out fifty repeats, how much mechanical work will she do in that time? [3200 J]

If it takes her 2 seconds per cycle (pull and release), at what fraction of a horsepower would she be expending energy? [at approximately 4.29% of one horsepower]

Exercise 12.b

A grindstone is operated by a person turning a hand crank, **A**, while a second person holds a tool against the grindstone to sharpen it. The crank radius is 200 mm. The shaft of the hand crank carries a fixed pulley **B**, which is connected to a pulley **C** by a drive belt. Pulley **C** is fixed on the shaft that carries the grindstone, **D**.

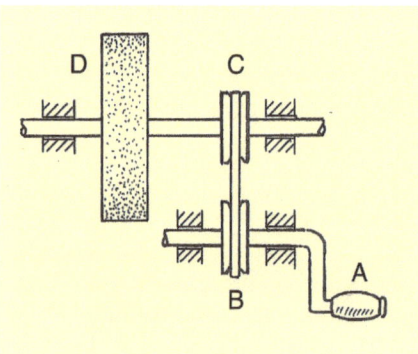

Available are:
Six pulleys, with respective diameters of 100 mm, 160 mm, 200 mm, 240 mm, 280 mm and 320 mm. Also, two grindstones, with diameters 300 and 500 mm respectively.

Which arrangement will result in the highest peripheral speed at the surface of the grindstone? State the required dimensions:

diameter of **B**: .... diameter of **C**: .... diameter of **D**..............

- Assuming the optimum values of the respective diameters has been chosen: if the operator can turn the crank at 3 rev/sec, what will be the linear speed of a point on the surface of the grindstone? [15.08 m/s]

  Supposing (in another scenario) that the diameters are: pulley **B**: 240 mm, **C**: 160 mm, and **D**: 600 mm: if a tool being sharpened on the surface of the stone results in a tangential friction force of 10 N, what magnitude of tangential force is needed on the crank handle? [22.5 N]

- If a person operates this grindstone in this way continuously for five minutes (while the tool is being sharpened), how much mechanical work will he or she have performed? [25.45 kJ]

- What fraction of a horsepower does this rate of work represent? [11.4%]

Exercise 12.c

A solid steel cylinder is firmly welded to a compression coil spring. This assembly is dropped from a height 'a' inside a long tube, so that it can bounce back up the tube. The height 'b' to which it rises after the bounce can be measured.

The mass of the cylinder is 6 kg, and that of the spring 2 kg. When tested separately, the coils of the spring close up against one another when a compressive force of 1248 N is applied to it. The natural length of the spring is 308 mm, and when it is completely compressed, its length is 116 mm.

Determine:

- The value of the spring constant. [6500 N/m]

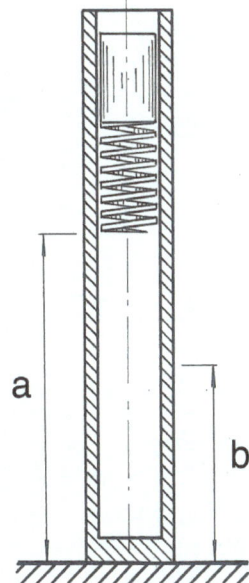

- The height 'a' that will give rise to 'b' being a maximum. [1.527 m]

Without performing calculations, make a reasoned estimate of what you expect the maximum bounce height 'b' to be, and explain why you expect it to have this approximate value.

Exercise 12.d

An assembly consisting of two plates, two coil springs and a pulling bracket, fits inside a square-section tube. It must be pulled upward inside the tube by a force P, which originates outside the tube.

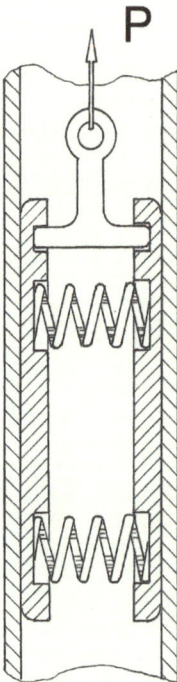

The mass of the moveable part of the assembly is 10 kg, the stiffness of each of the springs is 1200 N/m and the coefficient of friction between the plates and the inside of the tube is 0.3.

The uncompressed springs are each 80 mm long. They have to be compressed to a length of 50 mm to get the assembly into the tube.

Determine:

- The force in each spring when the assembly is inside the tube. [36 N]
- The amount of energy needed to compress the two springs. [1.08 J]
- The value of force P that will be just large enough to cause the assembly to slide upward inside the tube. [141.3 N]
- The work done by force P in moving this assembly up the tube by 600 mm. [84.78 J]

Exercise 12.e

A load of 4000 bricks, each of mass 3 kg, must be taken from point **A** to point **B**, a distance of 100 m up a small hill, inclined at 20° to the horizontal.

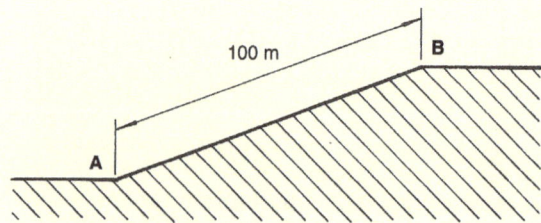

- How much mechanical work must be done on the bricks to accomplish this? [4026 MJ]
- If the bricks are transported at slow constant velocity by a 22 tonne truck with rolling resistance 5 kN, what driving force is needed at the wheels? [119.1 kN]
- If the output power to the wheels is 600 kW, how long will the trip take? [19.85 seconds]
- What would be the velocity of the truck while transporting this load? [18.14 km/h]

Exercise 12.f

A hand-cranked winding drum is used to raise a weight of 300 N slowly through a height of 4 m. The wire rope passes over a fixed, solid steel cylinder of diameter 120 mm. The drum diameter is 200 mm and the crank radius is 350 mm.

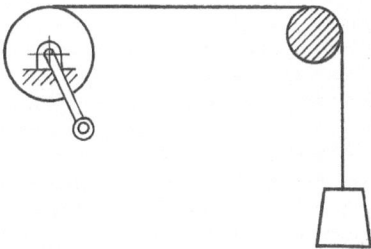

When a rope passes around a solid fixed cylinder, the ratio between the tensions in the tight side and the slack side of the rope is given by the equation:

$$T/S = e^{\mu\theta}$$

where e is the natural number 2.71828, also known as Euler's number, μ is the value of the coefficient of friction between the rope and the cylinder, and θ is the angle of contact between the rope and the cylinder, expressed in radians.

If $\mu = 0.4$ in this case, determine:

- The tension in the wire rope going onto the winding drum [562.3 N],
- The work done by the person turning the crank in raising the load through this height [2249 J],
- The percentage of this work that is useful work [53.35%],
- How long this process would take if the operator was capable of a power output of 20% of one horsepower [15 seconds], and
- The tangential force the operator would have to exert on the crank to prevent the load from slipping downwards. [45.73 N]

13

Simple lifting machines

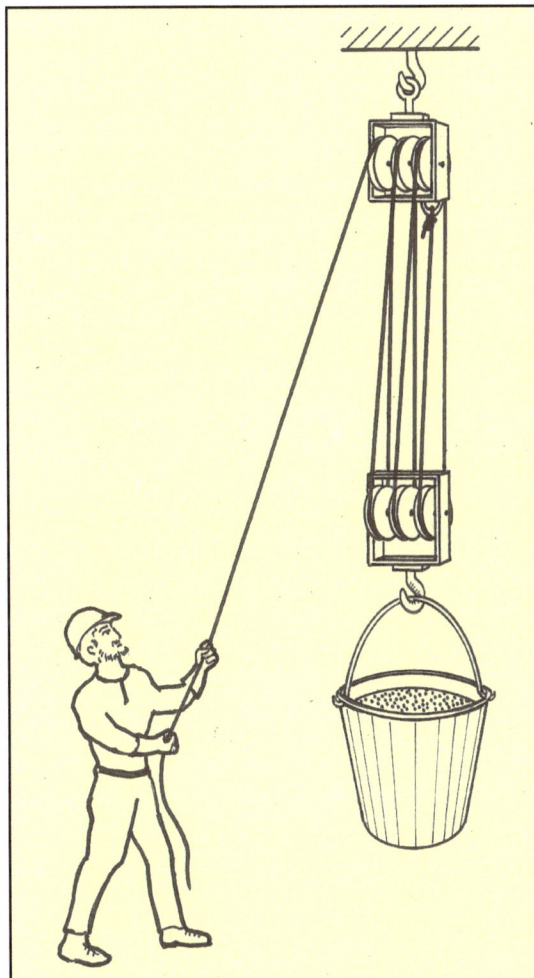

True/false tests on this topic

	True/false Test # 13a Simple Lifting machines			
1	A simple lifting machine (SLM) is designed to be powered by human muscular effort.	T	F	?
2	Simple lifting machines are entirely mechanical.	T	F	?
3	The value of the mechanical advantage of a SLM is never less than 1.	T	F	?
4	SLMs with very high values of mechanical advantage are likely to be impractical.	T	F	?
5	A SLM can enable a person to raise very large loads in the same time that it would take that person to raise a small load.	T	F	?
6	One can only get less mechanical work out of a SLM than was put into its operation.	T	F	?
7	In determining the velocity ratio of a SLM, the load distance can be measured in any direction that the load could be moved.	T	F	?
8	The distance moved by the effort is sometimes less than the distance moved by the load.	T	F	?
9	If there were no energy losses during the operation of a SLM, the mechanical advantage would have the same value as the velocity ratio.	T	F	?
10	The work done to raise machine parts is never recoverable.	T	F	?

	True/false Test # 13b Simple lifting machines			
1	The velocity ratio of a given SLM is always constant, irrespective of the load it needs to raise.	T	F	?
2	The energy required to stretch ropes in a block and tackle arrangement is recoverable.	T	F	?
3	The velocity ratio of any SLM can be deduced from the contributions to the overall velocity ratio made by the various parts of the machine.	T	F	?
4	A standing sheave (one whose axis is fixed relative to a machine frame) increases the velocity ratio by a factor of 2.	T	F	?
5	A running sheave (one that moves upwards during the raising of a load) increases the velocity ratio of a SLM by a factor of 2.	T	F	?
6	The diameter of any single sheave, whether standing or running, makes no difference to the velocity ratio of a machine.	T	F	?
7	It is possible to make up a block and tackle that uses only running sheaves.	T	F	?
8	In the commonest form of a block and tackle used for raising loads, called a 'second system', there is no practical limit to the number of sheaves in the upper block or the lower block.	T	F	?
9	A quick way of determining the velocity ratio of a 'second system' is to count the number of load ropes.	T	F	?
10	It is possible to set up a block and tackle of the type seen in a 'second system' that has a velocity ratio which is an odd number, like 3, 5 or 7.	T	F	?

	True/false Test # 13c Simple lifting machines			
1	The efficiency of a SLM is the ratio of useful work output to work input for a given operation of the machine.	T	F	?
2	The graph illustrating the variation of effort with load is generally a straight line graph.	T	F	?
3	The mechanical advantage of a SLM can sometimes be greater than the velocity ratio of that machine.	T	F	?
4	Any given SLM can be made more efficient, until its efficiency approaches 100%.	T	F	?
5	Whether or not a SLM will 'overhaul' (i.e. allow the load to descend when the effort is removed) depends on the amount of friction encountered in the operation of the machine.	T	F	?
6	A wheel and differential axle has four parts that move relative to the frame of the machine.	T	F	?
7	In calculating the velocity ratio of a lifting jack that relies on a nut moving up a screw thread on a post, the pitch of the screw thread is irrelevant.	T	F	?
8	A chain hoist could have a very large velocity ratio if the difference between the two diameters of the compound sheave is large.	T	F	?
9	In a block and tackle being used to raise a load, the tension in the rope is the same at all points.	T	F	?
10	When a wedge is used as a lifting device, it is possible to calculate the value of the effort force required to raise a given load.	T	F	?

Questions requiring short descriptive answers

1. Define what is meant by a simple lifting machine.
2. Name five basic components from which simple lifting machines can be constructed.
3. Name three simple lifting machines that can still be found in use today.
4. What would be wrong about claiming a certain simple lifting machine has a mechanical advantage of 0.6?
5. Provide a comprehensive definition of the velocity ratio of a simple lifting machine.
6. In connection with simple lifting machines: explain why the distance moved by the effort force will always be greater than the distance that the load is raised.
7. Explain why a standing sheave does not alter the velocity ratio of a simple lifting machine, while a running sheave does.
8. In a block and tackle, what are the respective effects of sheave diameter on the velocity ratio, the amount of friction caused, and the total energy loss?
9. Describe the steps in determining the velocity ratio for any given simple lifting machine.
10. Describe one quick way of determining the velocity ratio of a block and tackle of the 'second system' type, most commonly used for raising loads.
11. Explain why the efficiency of a simple lifting machine increases with load.
12. Why do some machines not 'overhaul' (i.e. allow the load to descend) if the effort force is accidentally removed?

Calculation exercises

Exercise 13.a

Suppose you need to design a block and tackle system for one man (assumed capable of exerting a force of approximately 200 N while pulling downwards) to raise a load of 1500 N through a height of 3 m. Assume an efficiency of 90%, and assume there is no stretch in the rope. The sheaves are 120 mm in diameter.

- What velocity ratio is required? [approximate = 7.5 so: say 8]
- How many sheaves would you need in the top block, and in the bottom block? [4 in each block]
- What is the minimum length of rope you would need, to be sure of having enough rope to hold at the start of the pull? [approx. 30 m]
- What will be the greatest tension in any part of the rope? [208.3 N in the effort rope]
- If the man can work at a rate of 300 W, how long would it take him to raise that load the full height? [approx. 17 sec]

Exercise 13.b

A motor car scissors jack is used to raise one side of a small car to

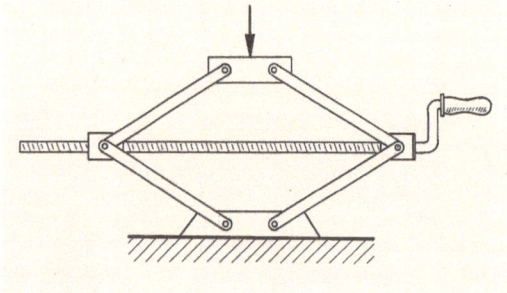

change a wheel. The mass to be raised is 560 kg, and the height by which it must be raised is 145 mm. Assume the efficiency of the jack to be 42%.

Draw an energy accounting diagram for this operation of the jack, to scale. Ignore the mass of the parts of the jack that need to be raised.

Using the diagram, determine:

- The work done by the person operating the jack. [1897 J]
- The amount of work done against friction. [1100 J]
- The time it would take to perform this operation, if the person could perform mechanical work at a rate of 75 watts. [about 26 sec]

Exercise 13.c

The diagram shows a plan view of a geared winch with the following dimensions: Crank handle radius 300 mm; gear attached to crank has 24 teeth; gear attached to drum has 96 teeth and drum diameter is 90 mm. Determine its velocity ratio. [26.67]

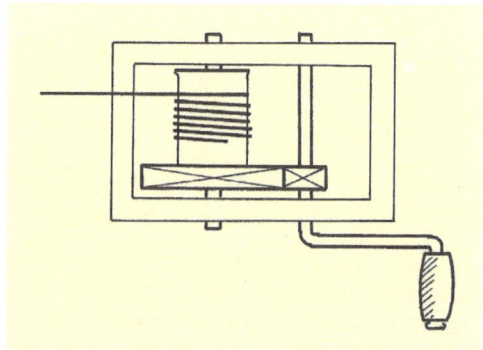

Exercise 13.d

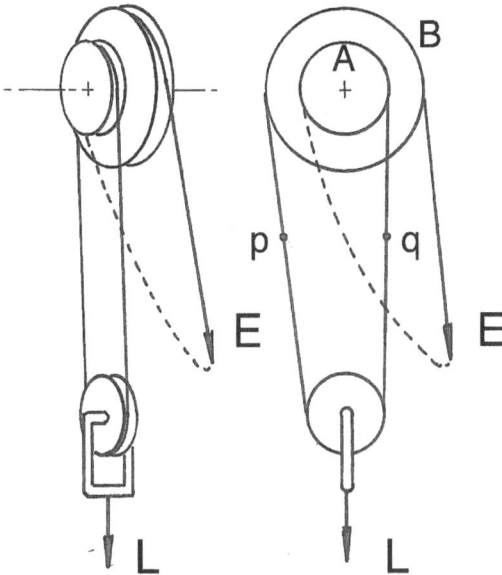

A differential chain block has a continuous chain passing over a standing double wheel, of which the larger side (**B**) has 40 links and the smaller side (**A**) has 36 links.

The running sheave in the loop of the chain has 16 links. The loop of chain is pulled by an operator exerting a downward effort force **E**. Determine the velocity ratio of this arrangement. [20]

Exercise 13.e

For the lifting arrangement shown here, the following data apply:

Angle $\theta = 40°$.
The weight of the rope is negligible.
Load box **A**
with sheave **B**
attached weighs
50 N.
Diameters of
sheaves **B** and **C**
are 100 and 200
mm respectively.
Compound drum
D has diameters
200 and 400 mm
respectively.

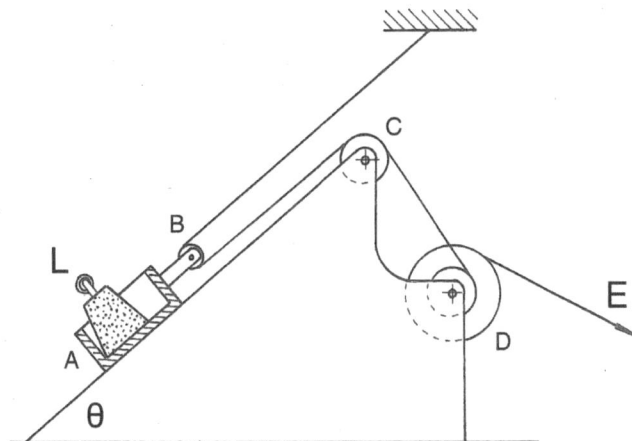

The effort rope makes an angle of 25° with the horizontal.
The coefficient of kinetic friction between load box and plane is 0.3

Reasoning from first principles, determine:

- The overall velocity ratio [6.223]

- For a load weighing 100 N: the effort required [32.72 N]; the mechanical advantage [3.056]; and the efficiency [49.11%]

- For a load weighing 200 N: the effort required [54.54N]; the mechanical advantage [3.667]; and the efficiency [58.93%]

- The equation describing the Law of this machine [E = 0.2182 L + 10.9]

Note: some of the given data are not needed in this exercise.

14

Inertia in linearly accelerating systems

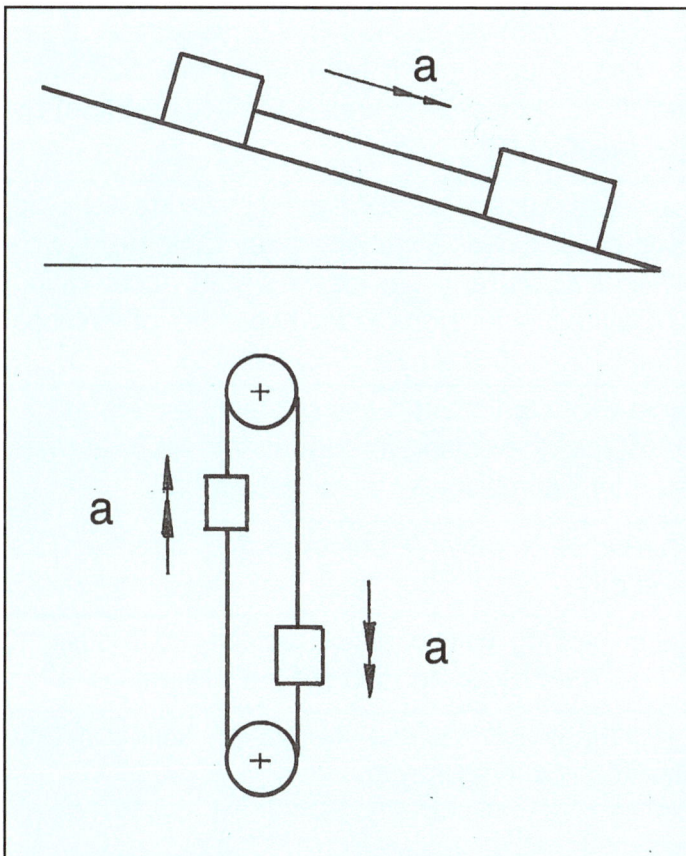

True/false test on this topic

	True/false Test # 14a Inertia in linearly accelerating systems			
1	Inertia could be thought of as that property of an object which makes it difficult to change the state of motion of the object.	T	F	?
2	Objects that are at rest don't possess inertia.	T	F	?
3	Newton's Second Law can only be applied to one mass with one force acting on it.	T	F	?
4	When a train of masses, linked by ropes, is made to accelerate by a force pulling the mass at one end of the train, the tensions in all the linking ropes have the same value.	T	F	?
5	For a system of linked masses, the acceleration of the system can be determined by dividing the total force (i.e. that force available to cause the acceleration) by the total mass of the system.	T	F	?
6	If two unequal masses are connected by an inextensible rope that passes over a stationary rotatable drum with insignificant mass, and the masses start accelerating due to gravity, the tensions in the two sides of the rope will differ.	T	F	?
7	A vehicle with significant mass can only accelerate to the extent that the driving force that acts on it remains greater than the resisting forces that act on it.	T	F	?
8	The inertia of a moving object is the same as its momentum.	T	F	?
9	An object moving with constant velocity in a straight line is governed by an equation of equilibrium.	T	F	?
10	An object that is moving in a straight line with constant acceleration is in equilibrium.	T	F	?

Questions requiring short descriptive answers

1. One can determine the acceleration of a train of linked objects by using the total force available and the total mass that has to be accelerated. However, doing it this way instead of investigating the motion of each of the linked objects, one at a time, will result in some information remaining unknown. Describe which information will be not be revealed, and why.
2. When using the equation $\Sigma F = ma$ for an object acted on by several forces, what determines the way that the directions of the forces are indicated?
3. When analysing the acceleration of a chain of linked objects, why is it necessary to assume that the ropes linking the objects do not stretch?
4. Explain what the term 'inertia' means.

Calculation exercises

Exercise 14.a

Two blocks of hardwood are joined by a light non-stretch rope passing over a light sheave with minimal friction in its bearings. The upper block rests on a steel surface, where the coefficient of kinetic friction between the two surfaces is 0.5.

Initially, the upper block is restrained from moving. When the upper block is released, the weight of the lower block causes the system to move.

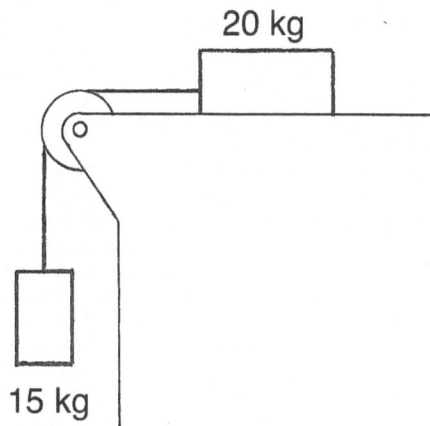

20 kg

15 kg

Determine the value of the resulting acceleration and the tension in the rope while accelerating. [1.401 m/s²; 126.1 N]

Exercise 14.b

Two mass-pieces are joined by a light inextensible rope that passes over a light sheave that is free to turn with negligible friction. The mass-pieces are initially restrained from moving. When they are released, the system starts to move.
Determine:

- The tension in the rope during the resulting movement. [490.3 N],
- How long it takes before the 49 kg mass-piece hits the stop. [7.139 seconds],
- The velocity of the 49 kg mass-piece when it hits the stop. [1.401 m/s] and
- The momentum of the system when it hits the stop. [140.1 N.s]

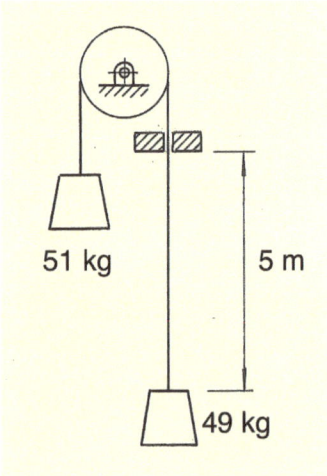

51 kg

5 m

49 kg

Exercise 14.c

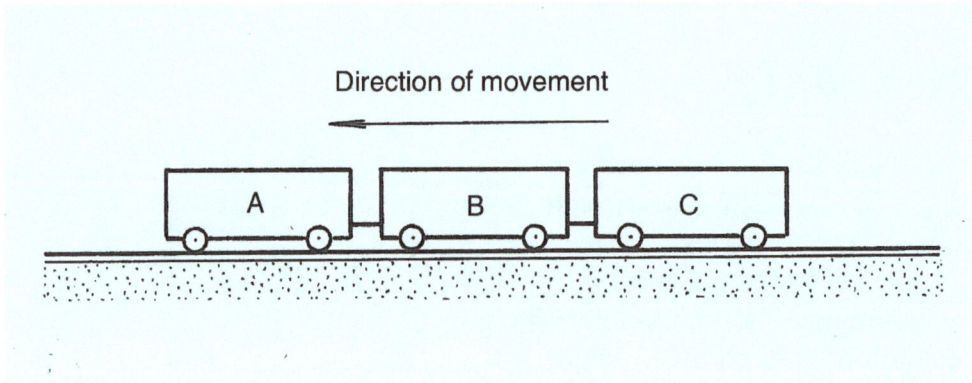

Direction of movement

A   B   C

A train of three linked railway wagons, each of mass 20 tonnes, is freewheeling on a level stretch of track, at 20 m/s. Each of the wagons experiences a rolling resistance of 3 kN and air drag of 2 kN at this speed. The emergency brake on the rear wagon is suddenly applied, providing a braking force of 100 kN.

Determine:

- The value of the deceleration of the train [1.75 m/s^2],
- The tension in the link between the rear and middle wagons [70 kN],
- The tension in the link between the front and middle wagons [40 kN]

- The distance in which the train will be brought to rest, if the air resistance remained constant. [114.3 m], and
- Whether the braking distance would be greater or less than this value, if air resistance varied with velocity, as it does in nature. [your call!]

Exercise 14.d

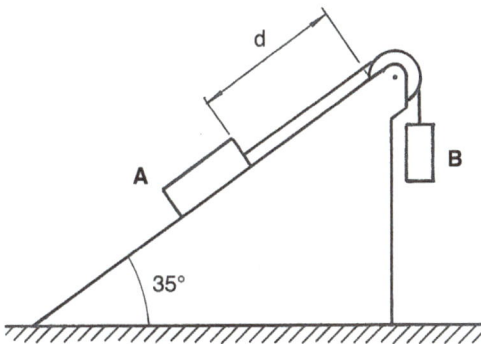

Block **A** (of mass 20 kg) rests on an inclined plane, where the coefficient of friction between the block and the plane is 0.35. This block is linked to block **B** by a light inextensible cord.

Assume the sheave has negligible mass and provides no frictional resistance to movement. Initially a brake prevents block **A** from moving. If dimension **d** is 900 mm, determine:

- The mass of **B** such that when the brake is released, the system accelerates to the right at 0.1 m/s^2. [17.59 kg] and
- Whether or not block **A** will crash into the sheave if the cord suddenly breaks exactly 4 seconds after the motion commences. [no]

Exercise 14.e

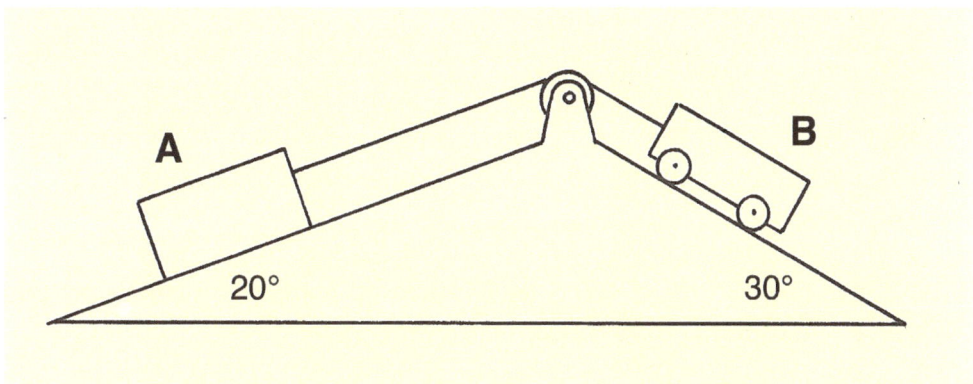

The coefficient of friction between block **A** (mass 20 kg) and the plane on which it rests is 0.3. Joined to block **A** by a light inextensible rope is trolley **B** (mass 5 kg), which is on a back-to-back incline. The sheave has negligible mass and may be assumed frictionless. The rolling resistance of the trolley is virtually zero.

- Prove by suitable calculations that the system would remain at rest if unconstrained.

- Determine how much mass needs to be placed in the trolley so that the system will accelerate to the right at 0.1 m/s². [9.381 kg]

Exercise 14.f

Two sturdy tables with machined horizontal surfaces are fixed to the floor. Wooden block **A** with rubber facing is suspended from two light inextensible cords attached to blocks **B** and **C** respectively. The cords pass over light sheaves that turn with negligible friction.

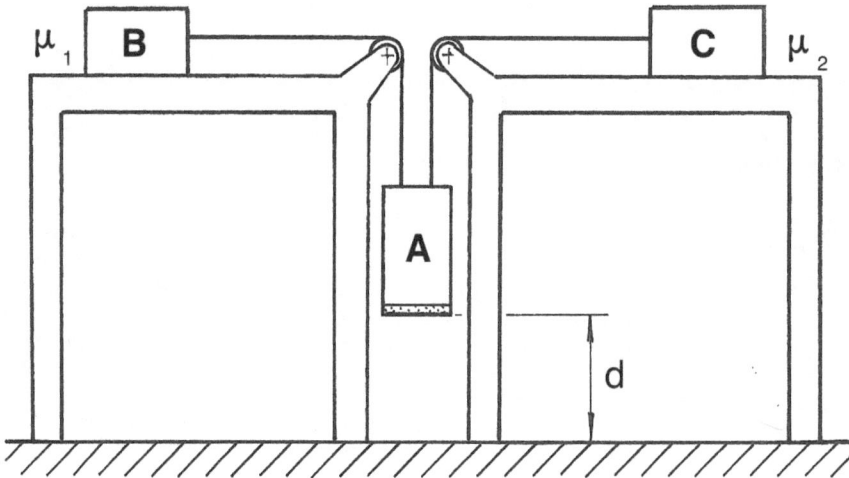

To begin with, blocks **B** and **C** are restrained from moving. When they are released simultaneously, block **A** descends to the floor, a distance **d**.

If the following data apply: Mass of **A** is 25 kg, mass of **B** is 20 kg, $\mu_1 = 0.3$ and $\mu_2 = 0.5$, **d** = 0.5 m and the time taken to descend this distance is exactly 0.5 seconds.

Determine the tensions in each cord and the mass of block **C**. [78.86 N; 141.4 N; 23.94 kg]

15

Linear momentum and impulse

True/false tests on this topic

	True/false Test # 15a **Linear momentum and impulse**			
1	The momentum of an object is the product of its mass and its speed.	T	F	?
2	Collisions between two objects can result in a change of momentum for one or both objects.	T	F	?
3	A billiard ball striking another billiard ball could be an example of an intentional momentum transfer.	T	F	?
4	A stream of fluid cannot be used to bring about a momentum transfer.	T	F	?
5	The Law of Conservation of Momentum states that the total momentum of the universe remains unchanged.	T	F	?
6	For the purpose of applying the Law of Conservation of Momentum, when a bat strikes a ball, the bat and the ball could be considered a closed system.	T	F	?
7	If two objects collide, they exert equal and opposite forces on one another for the duration of that interaction.	T	F	?
8	When two moving objects collide, if they rebound from one another, the object with the smaller mass will always rebound with a greater velocity than the other one.	T	F	?
9	The desk toy known as Newton's Cradle demonstrates the Law of Conservation of Momentum.	T	F	?
10	Firing a cannon-ball against an earth bank which brings the ball to rest, results in a net change to the momentum of the Earth.	T	F	?

	True/false Test # 15b **Linear momentum and impulse**			
1	In a collision between two objects, some energy can be lost but all the momentum is conserved.	T	F	?
2	The amount of momentum that is transferred from mass **A** to mass **B** in a collision depends only on the values of their respective masses.	T	F	?
3	A ballistic pendulum enables experimenters to determine the velocity of a high-speed object, by deducing what momentum change would have been necessary to make the pendulum swing to the measured height.	T	F	?
4	A change in momentum that occurs over a short finite period is called an impulse.	T	F	?
5	The coefficient of restitution is an index that describes the extent of the restoration of the shapes of two objects that have collided.	T	F	?
6	To analyse the forces that must have occurred in a collision, it is reasonable to estimate the peak force to be approximately twice the value of the average force experienced between the colliding objects.	T	F	?
7	If a hammer strikes a nail downwards into a block of wood, the forces that must be taken into account when determining the resulting impulse include the weight of the hammer and the weight of the nail.	T	F	?
8	It is impossible to determine the exact value of the peak of an impulsive force.	T	F	?
9	Suppose a bullet penetrates an earth bank behind a target, coming to rest: since the bullet no longer possesses momentum, this is an example that contradicts the law of conservation of momentum.	T	F	?
10	The units of an impulse are N.s (newton-seconds).	T	F	?

Questions requiring short descriptive answers

1. Define 'momentum' and provide three examples of situations in which large values of momentum may be present.
2. Provide three examples of situations in which transfers of momentum are caused intentionally.
3. Describe what the law of Conservation of Momentum states.
4. Under what circumstances should calculations be based on the Law of Conservation of Momentum, rather than on the Law of Conservation of Energy? Why?
5. How does the desk toy known as 'Newton's cradle' demonstrate the Law of Conservation of Momentum?
6. If momentum is always conserved, what happens to the momentum of a bullet that penetrates an earth bank behind a target?
7. Which principle of mechanics explains the phenomenon of 'recoil' that occurs when a bullet is fired from a rifle?
8. Explain the principle of operation of a ballistic pendulum.
9. Explain what is meant by an 'impulse' and an 'impulsive force' in relation to momentum.
10. Give reasons why it is not possible to predict how much energy will be lost in a collision, without empirical testing.
11. Describe what is meant by a 'coefficient of restitution' and state its uses.
12. Name some of the circumstances that would assist us to reduce the number of unknowns, when applying the Law of Conservation of Momentum to 2-D collisions.
13. Describe two examples where there is a deliberate momentum transfer making use of a stream of fluid.
14. Explain why a sailboat could not be made to move by directing a stream of air into the sail from an onboard fan powered by batteries.

Calculation exercises

Exercise 15.a

A steel sphere of density 7700 kg/m^3 and diameter 200 mm is released from rest and rolls down a slope where it strikes a large nail that is partially embedded in a wooden post. The mass of the nail is 400 g.

On being struck, the nail penetrates an additional 156 mm into the post.

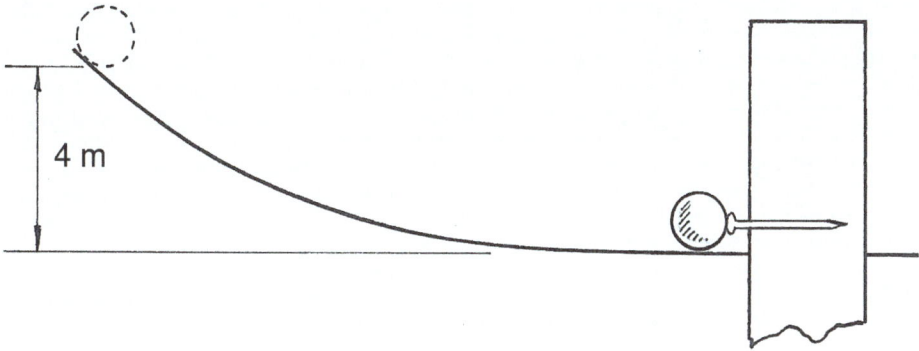

Determine:

- The mass of the ball [32.25 kg],
- The velocity of the ball immediately before striking the nail [8.859 m/s],
- The velocity with which the ball and the nail start moving after the collision [8.750 m/s],
- The value of the deceleration of the nail in coming to rest [245.4 m/s^2]
- The average resisting force that the wood offers to penetration by the nail. [8013 N]
- The amount of energy lost in the collision [15.50 J]

What happened to the rest of the energy? Explain.

Exercise 15.b

An arrow of mass 60 g is fired horizontally into a 10 kg target that consists of a block of expanded polystyrene to which is glued two wooden slabs. This target rests on a flat horizontal table where the coefficient of kinetic friction between the wood and the table is 0.4. On impact, the arrow becomes imbedded in the polystyrene, and the target is found to move a distance d = 10 mm.

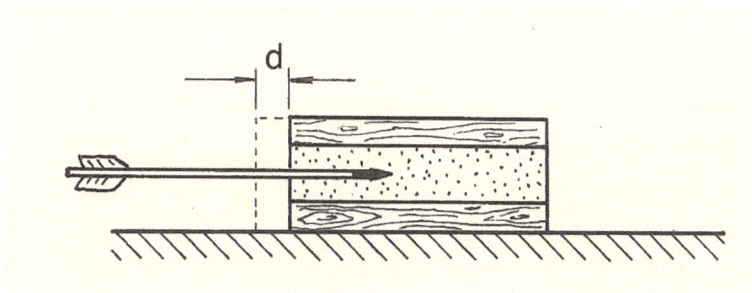

Determine:

- The amount of work done against friction with the table while the target is sliding. [0.3948 J]

- The velocity of the target immediately after impact, if it is assumed that 80% of the arrow's original energy before impact is lost due to a combination of work done to deform material, and the production of heat and sound. [0.6264 m/s]

- The velocity of the arrow on impact. [105.0 m/s]

Exercise 15.c

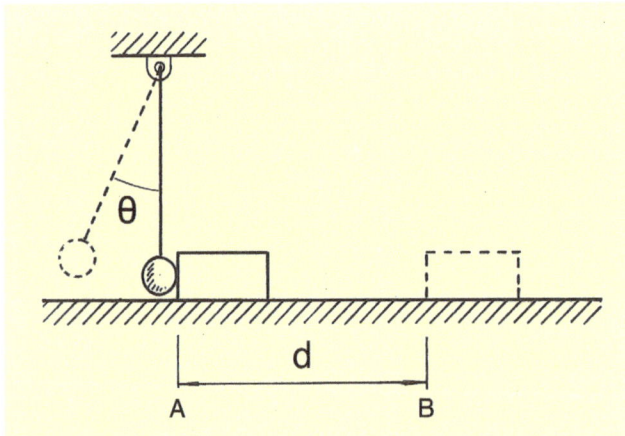

A light rod with a spherical mass-piece of exactly one kg attached firmly to one end, is hinged to be able to swing from an overhead pivot, with negligible friction. From the pivot axis to the centre of the sphere is 800 mm.

The rod is raised to make an angle θ with the vertical, then released to swing down to strike squarely a stationary block of mass 400g, resting on a level polished table surface, where the coefficient of kinetic friction between the block and the table is 0.1.

After the collision, the sphere continues in the same direction with a velocity of 10% of its velocity before the collision. Determine the value of angle θ such that the block will slide a distance of exactly one metre before stopping at point B. [12.76°]

Exercise 15.d

A pile driver of mass 3.2 tonnes falls from rest, from a height h = 3.8 m, onto a 120 kg steel stake, to drive it into the ground. If the stake penetrates the ground by 450 mm as a result of this impact, determine:

The velocity at which the stake begins penetrating the ground immediately after the impact. [8.322 m/s]

The average value of the deceleration of the stake while penetrating the ground. [76.96 m/s^2]

The average resisting force offered by the ground. [288.1 kN], and

The impulse of the strike. [998.6 N.s]

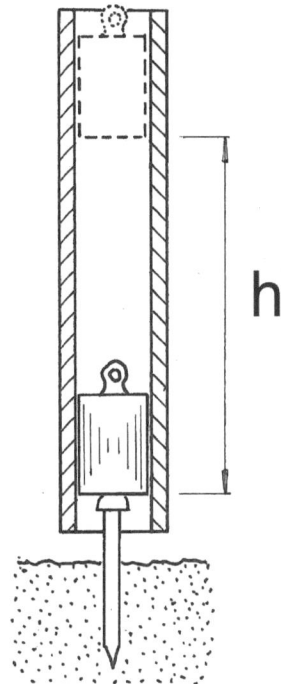

Exercise 15.e

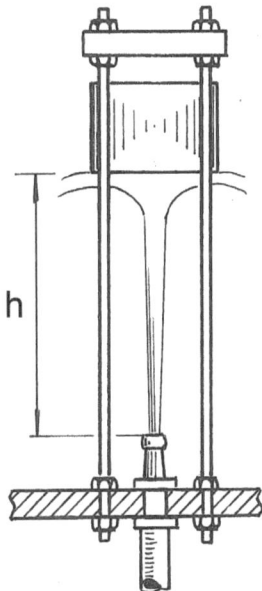

A stream of water emerging from a fire hose nozzle can keep a hollow metal cylinder at a steady height **h**, trapped inside four rods, within which it is free to slide without resistance.

If the following data apply:

The hollow cylinder weighs 100 N,
Height **h** is 3 m, and
The diameter of the nozzle mouth is 25 mm:

Determine the speed of the water emerging from the nozzle. [15.34 m/s]

Exercise 15.f

Explain the reasoning that enables us to determine the force produced by the steady momentum transfer due to a stream of fluid impinging on an obstacle. [*We determine the force it requires to change the momentum of the fluid stream, and deduce that the fluid stream must be exerting an equal and opposite force on the obstacle.*]

If an experimental jet engine on a test bench takes in still air, and forces the air out at a velocity of 400 m/s, producing thrust measured at 20 kN:

- What would be the mass flowrate of the air passing though this engine? [50 kg/s]
- If the density of the still air is 1.25 kg/m³, how many litres of still air is being forced through the engine every second? [40 × 10³ litres]

16

Relative velocity

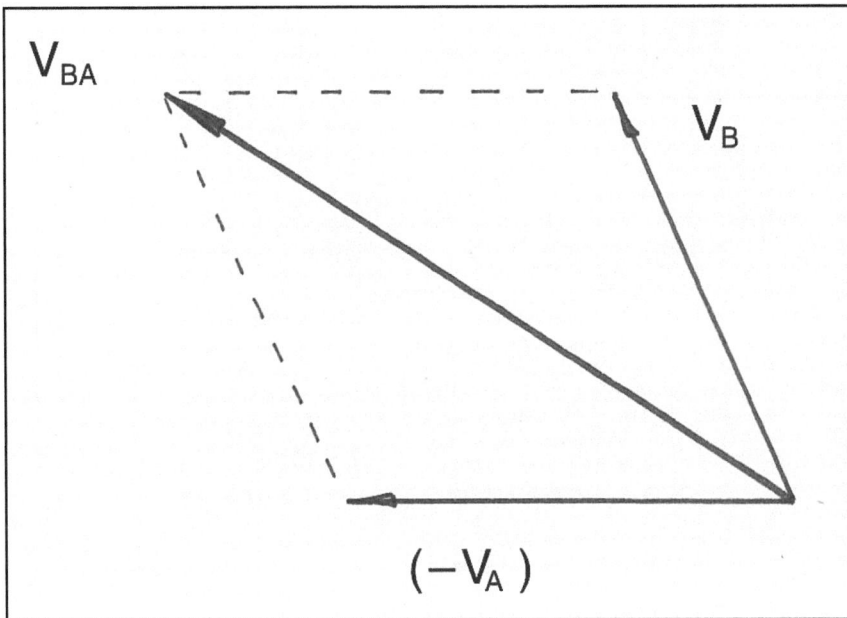

True/false tests on this topic

	True/false Test # 16a Relative velocity			
1	The 'true' velocity of an object is able to be defined without reference to any other object.	T	F	?
2	An 'absolute' velocity is conventionally taken to mean velocity relative to Earth.	T	F	?
3	If an object is moving in a field which is itself moving relative to Earth, then its absolute velocity can be found by a vector subtraction.	T	F	?
4	If you try to swim in a direct line across a flowing river, at a right angle to the banks, you will need to be able to swim faster than the speed at which the river is flowing.	T	F	?
5	The groundspeed of an aircraft that is flying is the vector sum of the airspeed and the windspeed.	T	F	?
6	There are three different conventions for describing the direction of a velocity vector in a given plane.	T	F	?
7	The relative velocity of moving vehicle **B**, as seen from moving vehicle **A**, is found by a vector subtraction.	T	F	?
8	The time it takes for two objects moving in the same plane to reach the point of closest approach is determined using their relative velocity and their relative path.	T	F	?
9	Suppose moving object **A** observes moving object **B**, both moving in a plane parallel to the Earth's surface: then the path on which **B** appears to be moving, relative to **A**, is a real path that can be traced on the ground.	T	F	?
10	The shortest distance that will occur between two objects moving in the same plane, is measured at right angles to their relative path.	T	F	?

	True/false Test # 16b Relative velocity			
1	The relative path between two moving objects always lies in the same direction as the relative velocity between them.	T	F	?
2	The condition for moving object **A** to intercept moving object **B** is that the direction of V_{BA} has to lie in the opposite direction to the line **BA**.	T	F	?
3	An archer shooting at a target that is moving on a straight path situated some distance from himself, will need to launch the arrow before the moving target reaches the point of closest approach.	T	F	?
4	In a mechanism, the instantaneous velocity of a slider relative to the rod that it slides on, has to lie in the same direction as the rod is pointing at that instant.	T	F	?
5	Abrasive material being poured onto a conveyor belt should preferably have a velocity relative to the belt, that is at right angles to the direction of movement of the belt.	T	F	?
6	For a point P on a pivoting rod that is rotating, at any given instant, the velocity of P relative to the pivot must be directed in the same direction that the rod is pointing.	T	F	?
7	In a mechanism consisting of a crank, connecting rod and piston, all of whose dimensions are known, it is possible to determine the velocity of the piston at any point of its motion, provided the angular velocity of the crank is known.	T	F	?
8	To subtract one vector, V_1 from another, V_2, is the same as adding the opposite vector, namely $V_2 + (-V_1)$.	T	F	?
9	The relative path of the motion of one object as seen by another, is a real path that can be traced on the ground.	T	F	?
10	For any two objects, **A** and **B**, that are moving in the same plane, the velocities V_{BA} and V_{AB} are equal and opposite.	T	F	?

Questions requiring short descriptive answers

1. Is there such a thing as an absolute velocity? Explain.
2. Suppose you are in the passenger seat of an enclosed car moving at 100 km/h, and you idly toss an orange up a short distance (say, level with your chin) and catch it. Why does the orange move in exactly the same way as it would if you did this when the car was stationary?
3. Describe what is meant by a 'resultant velocity', and give one example.
4. Describe two conventions for defining the direction of a velocity vector.
5. What is meant by a 'relative path'? Give one example.
6. A condition for there to be a collision between two objects moving on straight paths in the same plane, is that the relative path between them should...what?
7. Describe the procedure for determining the shortest distance that will occur between two objects moving on straight paths in the same plane.

Calculation exercises

Exercise 16.a

An aircraft with cruising speed 160 km/h needs to fly from **A** to **B**, which is a distance of 180 km due east.

a. If there is no wind, how long will the journey take? [1:07:30]
b. If a 60 km/h wind is blowing from the north-east, in what direction should the pilot head, in order to travel in a straight line from **A** to **B**? Draw a velocity diagram to scale, to illustrate how your answer is achieved. [E 15.38° N]
c. State this direction in the form of a bearing. [74.62°]
d. What will be the ground-speed of the aircraft? [111.8 km/h]
e. How long will the journey take? [1:36:34]

Exercise 16.b

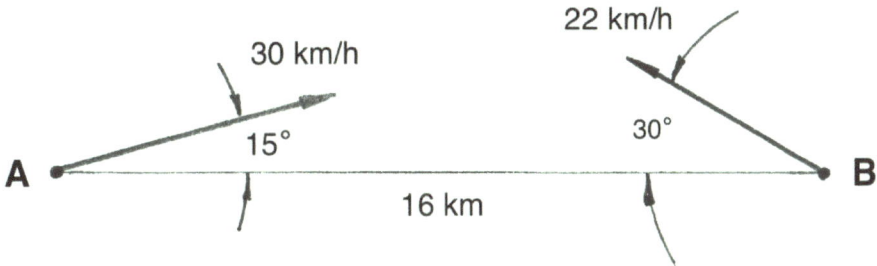

At a given moment, two ships are in the positions shown, with the velocities shown. If they both continue to hold their courses without changing velocity:

Draw a velocity diagram to scale 5 mm = 1 km/h and use it to determine the magnitude of V_{BA} in km/h. [48.14 km/h]

State the direction of V_{BA}, in the form of a bearing. [274.2°]

What will be the shortest distance between the two ships as they pass one another? [1.075 km]

How long will it take to get to the point of closest approach?
[19 min 54 sec]

Exercise 16.c

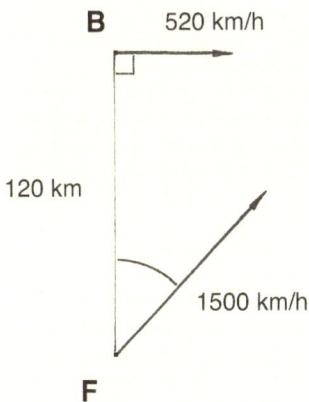

At a given moment, a fighter pilot whose aircraft is already airborne is alerted to the presence of an enemy bomber which is 120 km due north of him, and is flying at 520 km/h due west.

The maximum speed of the fighter is 1500 km/h.

The pilot is instructed to intercept the bomber in the shortest possible time.

Determine:

- The direction in which the fighter should head. State this direction in the form of a bearing. [20.28°]

111

- The magnitude of V_{FB}, in km/h [1407 km/h] and
- The time between sighting the bomber and interception [5 min 7 sec].

Exercise 16.d

The diagram shows the positions and velocities of two aircraft flying at the same altitude at exactly 10:00:00.

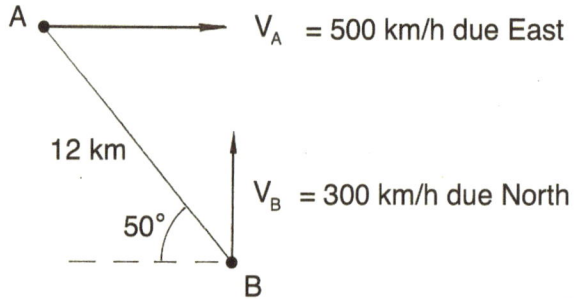

A V_A = 500 km/h due East

12 km

V_B = 300 km/h due North

50°

B

a. Determine the velocity of **A** relative to **B**. [583.1 km/h, on a bearing of 120.1°]
b. Draw on a map that replicates the accompanying diagram, the relative path of **A**, as seen from **B**.
c. If they continue flying on their original paths, what will be the closest distance between the two aircraft? [3.915 km]
d. At what time will they reach the point of closest approach? [10:01:10]

Exercise 16.e

In an ore-processing plant, crushed gravel emerges from a chute, landing on an inclined conveyor belt in a stream that makes an angle of 60° with the horizontal at the point of incidence.

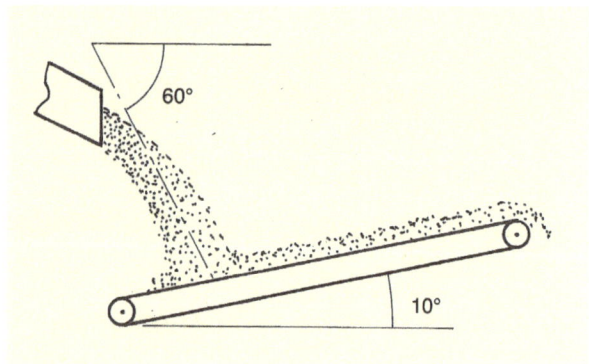

60°

10°

The incident velocity is 6 m/s. The belt is angled at 10° to the horizontal.

If the relative velocity of the gravel to the belt at the point of incidence should be at right angles to the belt, to minimise abrasion, determine a suitable speed for the belt. [2.052 m/s]

17

Centripetal and centrifugal forces

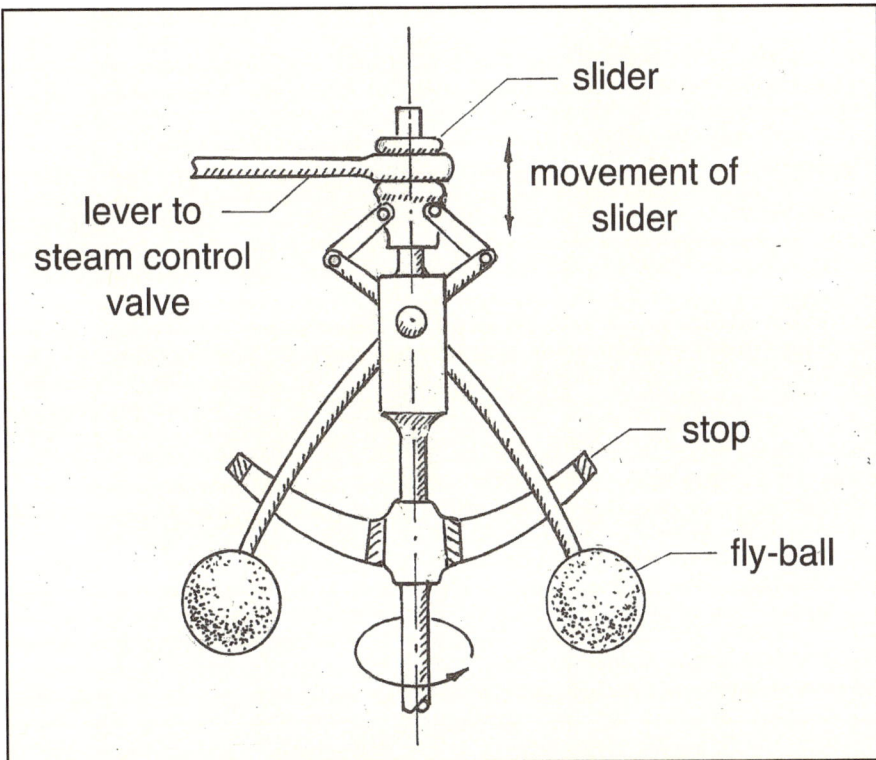

Steam engine speed governor designed by James Watt

True/false tests on this topic

	True/false Test # 17a Centripetal and centrifugal forces	T	F	?
1	In order to keep an object moving in a circular path, there has to be a force pulling the object towards the centre of the circle.	T	F	?
2	Centrifugal force is a type of inertial force.	T	F	?
3	It is possible for the centrifugal force on an object moving in a circular path to be greater than the centripetal force acting on it.	T	F	?
4	If the centripetal force acting on an object is suddenly removed, the object will continue moving in a straight line path, radial to the circle in which it had been moving.	T	F	?
5	The centripetal force on a vehicle that is cornering can be provided by friction between the tyres and the road.	T	F	?
6	When a fixed-wing aircraft has to change direction, the centripetal force is provided by the airstream over the tail fin.	T	F	?
7	A vehicle experiences an increase in the road reaction force when driving through a dip.	T	F	?
8	Centripetal acceleration is the rate of change of velocity towards the centre of the circular movement.	T	F	?
9	If a vehicle drives over a rise with a curved profile and is on the point of leaving the road, it is because at that moment, the centrifugal force is about to exceed the weight of the vehicle.	T	F	?
10	When an object is being constrained to move in a circular path, there is no force external to the object, that is pushing it radially outwards.	T	F	?

	True/false Test # 17b Centripetal and centrifugal forces	T	F	?
1	The reason why a fixed-wing aircraft has to bank In order to make a horizontal turn, is that a component of the lift force on the wings is needed to provide centripetal force.	T	F	?
2	It is possible for a cyclist to ride around a circular bend without leaning, if he or she rides slowly enough.	T	F	?
3	The force that is needed to keep a vehicle from sliding on a bend on a banked curve, can be provided by friction or gravity, or a combination of both.	T	F	?
4	A vehicle could overturn on a horizontal bend if it is unstable in a vertical plane that is radial to the bend.	T	F	?
5	An object moving in a curved path that is not properly circular, does not experience centrifugal force.	T	F	?
6	The forces to which fairground rides expose patrons, so that they experience unaccustomed sensations, are usually centrifugal in nature.	T	F	?
7	Mechanical speed governors used on early steam engines relied on flyweights moving outwards against a restraining force, due to centrifugal force.	T	F	?
8	The magnitude of the centrifugal force on an out-of-balance shaft rotating at high speed depends only on the mass of the shaft and the speed of rotation.	T	F	?
9	When analysing the forces acting on a four-wheeled vehicle negotiating a banked curve in a horizontal plane, if both the inner and the outer wheels are touching the road surface, the combined sideways friction force on the wheels may be considered as one force.	T	F	?
10	For a satellite to stay in a stable orbit around the Earth, it has to have the correct velocity for the altitude at which it is moving.	T	F	?

Questions requiring short descriptive answers

1. Explain the difference between centrifugal and centripetal force.
2. Why is it not possible to ride a bicycle around a curve on a horizontal road while rider and bicycle remain vertical?
3. Is the centrifugal force a real force? Explain.
4. If an object is constrained to move in a circular path with constant angular velocity, is it accelerating?
5. When a motor vehicle drives through a dip with a circular profile, what additional force does it experience, and what are the dangers of negotiating the dip at too great a speed?
6. When a motor vehicle drives over a rise in the road with a circular profile, if the radius of that profile is not great enough, what are the risks for the vehicle?
7. What provides the centripetal force that allows an airborne fixed-wing aircraft to make a turn?
8. When looping the loop in an aerobatic plane, what are the two respective circumstances that could cause the pilot to lose consciousness at the top and the bottom of the loop? Explain.
9. In order for a vehicle not to slide outward on a banked curve, a sufficiently large centripetal force is required. Which forces contribute to providing the required centripetal force?
10. The attraction of fairground rides is to subject passengers to unusually large forces to give them a thrill. This often involves causing them to experience circular motion at speed. What factor limits the size of the forces to which they should be exposed, for their own safety?
11. Describe the essential features of a mechanical speed governor.
12. If a rotating part of a machine is not properly balanced, a harmful vibration can result. Explain what causes this.

Calculation exercises

Exercise 17.a

Vertical shaft **ABD** can be rotated by a drive belt applied to the

pulley. To point **A** is attached a stirrup connected to a light wire, at the end of which is fixed a steel sphere, **C**.
AB = 500 mm, **AC** = 300 mm and the mass of the sphere is 4 kg.

Determine:

The speed, ω, at which the shaft should be rotated so that angle θ becomes 75°.
[11.24 rad/s]

The tension in wire **AC** at this speed.
[151.6 N]

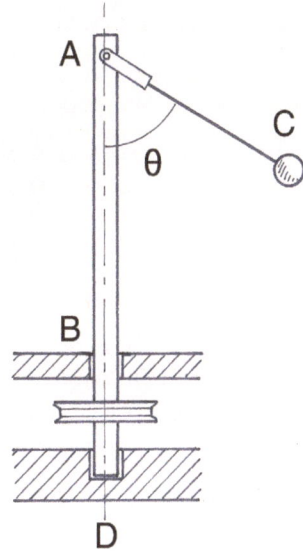

Exercise 17.b

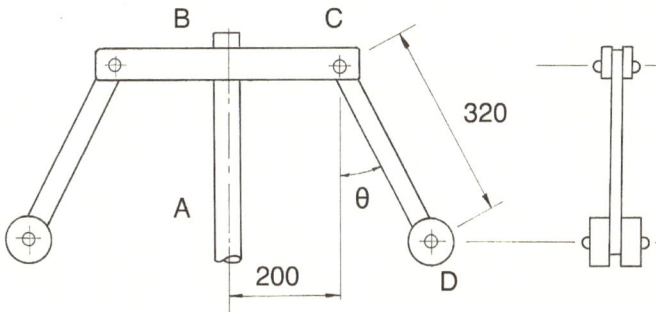

Rotating vertical shaft **AB** has two horizontal bars fixed to it. At each end of the bars is attached a bar **CD**, free to turn on the hinge pin at **C**. At the lower end of each rod are attached two discs. The two discs together have mass 4 kg, and each bar **CD** also has mass 4 kg.

Draw a free-body diagram of bar **CD**, showing all forces resolved into vertical and horizontal components.
Determine the rotational speed (in r/min) at which angle θ will be 30°.
[39.36 r/min]

Exercise 17.c

A motor car of mass 1426 kg and wheel-track width 1500 mm has its centre of gravity midway between the left and right wheels and 600 mm above the ground. This vehicle is driven around a curve in the road, of radius 80 m. The road surface is banked at 15° to the horizontal.

- Draw, to scale, a cross-section of the banked road with the vehicle on it, showing the contact points of the left and right wheels, and the centre of gravity of the car.
- Determine the speed at which the car would overturn on this road curve, assuming there is sufficient friction for it not to slide first. A graphical solution is permitted. [152.4 km/h]

Exercise 17.d

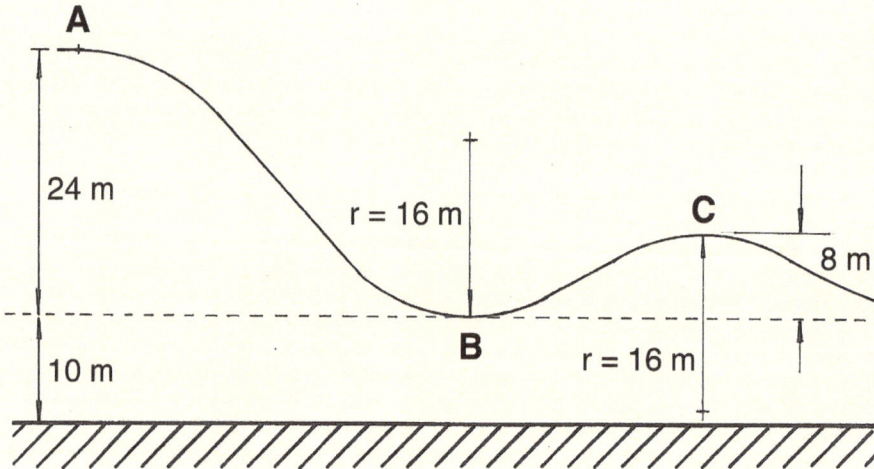

A rollercoaster car starts from rest at point **A** and moves, driven only by gravity, through a dip at point **B** and over a rise at point **C**. The mass of the car plus riders is 1200 kg. Assume there are no energy losses due to friction, and determine:

- The speed of the car at points **B** and **C** respectively. [21.70 m/s and 17.72 m/s]
- The thrust of the track on the wheels at points **B** and **C** respectively, [47.09 kN upward at point **B**; however, at point **C** a downward pull

118

of 11.77 kN is needed to keep the car in contact with the track] and
- The number of g's experienced by riders at points **B** and **C** respectively. [4 g's downward and 1 g upward respectively]

Exercise 17.e

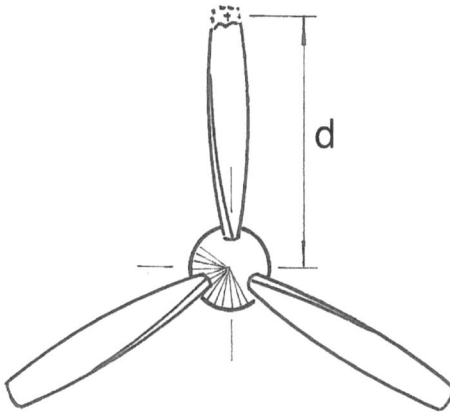

A WW2 fighter plane sustains damage to its three-bladed propeller in a dogfight. A small piece gets shot off the end of one of the propeller blades. The mass of this missing piece is 3 kg, and its centre of mass was d = 1.45 m from the axis of rotation of the propeller shaft.

Given that the mass of the intact propeller was 168 kg, determine:

- The distance that the centre of mass of the propeller has shifted due to this damage, [26.36 mm] and
- The consequent out-of-balance force on the propeller shaft bearings at a rotational speed of 1320 r/min. [83.12 kN]

18

Rotational inertia

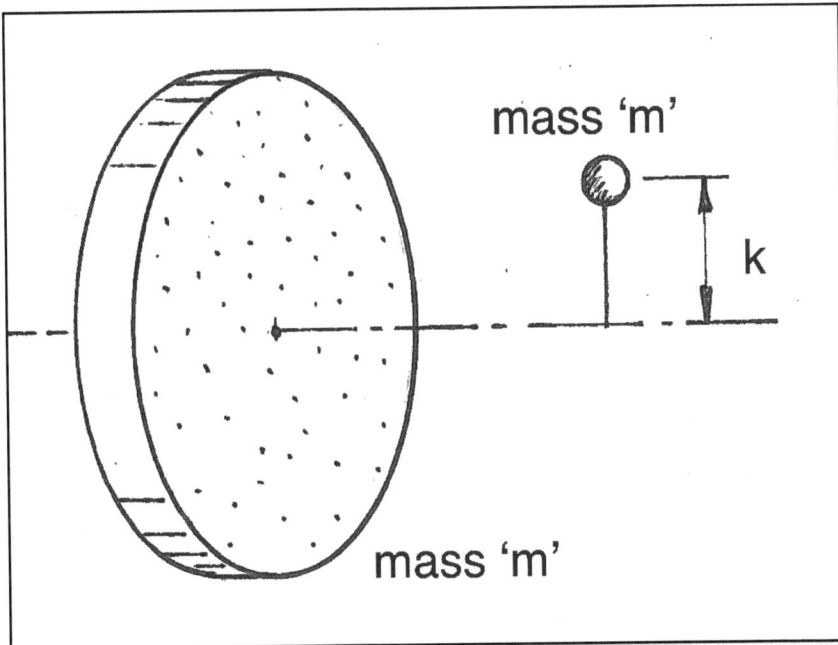

mass 'm'

k

mass 'm'

True/false tests on this topic

	True/false Test # 18a Rotational inertia			
1	A body's rotational inertia is directly proportional to the distance between its centre of mass and the axis about which the attempted rotation is occurring.	T	F	?
2	The radius of gyration of an object cannot be larger than the radius to the point on the object that is furthest from its axis of rotation.	T	F	?
3	The mass moment of inertia of a body can be ignored in calculations that apply to its motion while it is rotating with constant angular velocity.	T	F	?
4	The rate of angular acceleration of a body being rotated depends on its mass moment of inertia and the torque applied to it.	T	F	?
5	The radius of gyration of a body is the distance between the axis about which it is made to rotate, and the axis parallel to that one, that passes through the body's centre of mass.	T	F	?
6	The mass moment of inertia of a body about a given axis remains constant under all circumstances, provided the body doesn't change its shape.	T	F	?
7	The rotational inertia of an object about a given axis can only be established by calculation, not by experiment.	T	F	?
8	The possible values of mass moment of inertia for man-made objects that are designed to rotate could range from 10^{-10} to 10^{10} kg.m^2.	T	F	?
9	The rotational inertia of a thin rod rotated about an axis perpendicular to the rod, passing through its midpoint, is one third of the value of its rotational inertia when the rod is rotated about an axis parallel to that one, that passes through one end of the rod.	T	F	?
10	The radius of gyration of a rigid rotatable object is always constant, irrespective of the axis about which it is made to rotate.	T	F	?

	True/false Test # 18b Rotational inertia	T	F	?
1	The mass moment of inertia about a given axis, of a body consisting of a number of regular geometric shapes, can be determined by adding the MMI values of the individual shapes with respect to that same axis.	T	F	?
2	The formula for the MMI value of a thin-walled cylinder of mass **m** about its longitudinal axis of symmetry is the same as that for a point mass **m**, situated at twice the radius of the cylinder from that axis.	T	F	?
3	If a disc rotates about an axis that passes through its centre of mass, in the plane of the disc, its MMI value about that axis is twice that of the MMI value when it rotates about the axis that passes through the centre of mass at right angles to the plane of the disc.	T	F	?
4	The parallel axis theorem can be used directly to determine the rotational inertia of an object about any given axis of rotation, provided that its rotational inertia about an axis parallel to the given one is known.	T	F	?
5	The perpendicular axis theorem allows us to determine the MMI of an object about an axis that is perpendicular to any another axis about which the object's MMI is known.	T	F	?
6	The perpendicular axis theorem applies only to thin plates, not to thick solid objects.	T	F	?
7	When estimating the value of the MMI of a wheel with a hub, spokes, and a rim, the value for the whole wheel is likely to be only very slightly more than that of the rim.	T	F	?
8	It is possible to determine the MMI value of an object by a process that makes use of subjecting the object to slow rotational oscillations, and by measuring the period of those oscillations.	T	F	?
9	It is possible to determine the value of the MMI of some objects by allowing them to swing as a compound pendulum and measuring the period of oscillation.	T	F	?
10	The most reliable way to determine the MMI of a regular solid object is by a calculation based on the mass distribution of that object about the axis of oscillation.	T	F	?

Questions requiring short descriptive answers

1. Describe what the rotational inertia of an object is, and the factors that contribute to its magnitude.

2. Explain what is meant by a 'radius of gyration'.

3. Why is it essential to specify a value of Mass Moment of Inertia (MMI) with respect to a particular axis?

4. Explain why a thin-walled circular cylinder of radius 'r' has the same MMI value with respect to its axis of symmetry, as does a particle of the same mass situated a distance 'r' from that axis.

5. What does the Parallel Axis Theorem enable us to do?

6. What does the Perpendicular Axis Theorem enable us to do?

7. Describe a situation in which the Parallel Axis Theorem and the Perpendicular Axis theorem both need to be used.

8. Is the Rotational Inertia of an object equally of importance when the object is rotating at constant speed as when it is accelerating or decelerating? Explain.

9. Suppose you do a practical experiment to determine the value of the MMI about a given axis for an object that is a regular shape, such as a rectangular flat plate of homogeneous density, and obtain the value 1.75 kg.m². You then calculate this value based on the mass and dimensions of the object, and obtain the value 1.78 kg.m². Which value would you regard as more reliable, and why?

10. Describe the method of determining the value of the rotational inertia of an object by making use of rotational oscillations.

11. Describe the method of determining the value of the rotational inertia of an object by using it as a compound pendulum.

12. Provide a reasoned estimate of the likely value of the MMI of a small gearwheel in a watch.

13. With the aid of a sketch, show how one could estimate the MMI value of an impeller vane that is rectangular, but is not radial to the axis of rotation of the impeller.

Calculation exercises

Exercise 18.a

How many concentrated mass-pieces of 4 kg each need to be arranged at a radius of 800 mm to result in a mass moment of inertia of *at least* 40 kg.m²? [16 pieces]

When the assembly is put together and subjected to a torque of 50 Nm, what will be the resulting angular acceleration? Ignore the mass of the structure that supports these mass-pieces. [1.221 rad/s²]

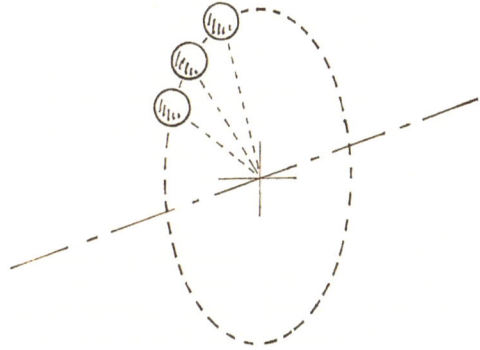

Exercise 18.b

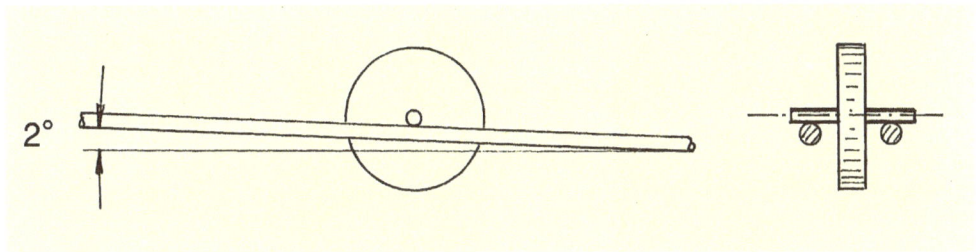

A solid disc with a fixed axle rolls down a pair of parallel rails, from a standing start, without slipping. The rails are inclined at 2° to the horizontal.

The mass of the disc plus axle is 20.00 kg. Its diameter is 320 mm. The diameter of the axle is 12 mm.

Determine the time that this disc should take to roll a distance of 800 mm along this pair of rails.

State clearly the principles you are using to solve the problem. [40.82 sec]

Exercise 18.c

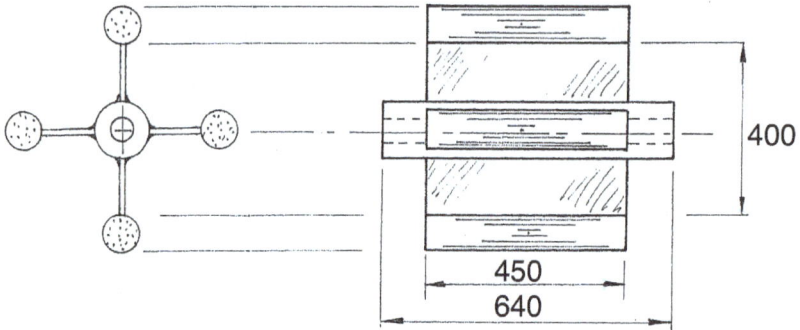

A rotating assembly is made of a central hollow steel cylinder, inside diameter 40 mm, and outside diameter 120 mm, with four flat steel plates, thickness 8 mm, welded to it. At the end of each plate is attached a solid copper cylinder, diameter 90 mm, and density 8900 kg/m³.

Determine the mass moment of inertia of the assembly about its axis of rotation. Consider the plates to be thin plates. Take the density of the steel to be 7800 kg/m³. [6.612 kg.m²]

Exercise 18.d

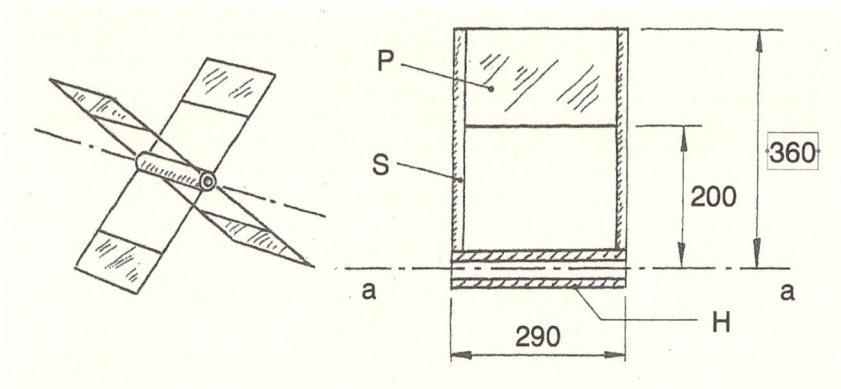

A paddle-wheel consists of a central hub **H**, to which four paddles **P** are joined by spokes **S**. All parts are made of brass, density 8400 kg/m³.
Hub **H** is of hollow circular section, OD 60 mm and ID 30 mm. Spokes **S** are made of solid round rod, diameter 20 mm. The paddles are 5 mm thick.

Determine the mass of **H**, one part **S** and one part **P**, respectively, accurate to the nearest gram. [5.166 kg; 871 g; 1.680 kg]

Using these values, determine the mass moment of inertia of the assembly about axis **a-a**, as accurately as possible, to four significant figures. [0.8723 kg.m²]

Exercise 18.e

A rectangular block of steel, 400 x 400 x 500 mm high, has a hole of diameter 300 mm bored through it.

The centre-line of this hole passes through the centres of the top and bottom faces of the block. This steel has density 7800 kg/m³.

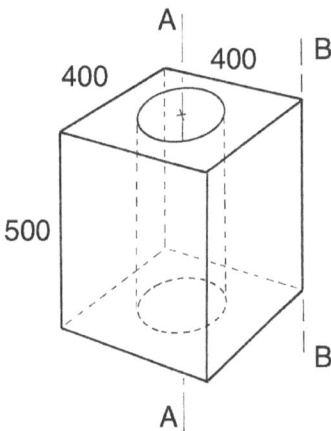

Determine:

*   The mass of the bored-out block. [348.3 kg]
*   The mass moment of inertia of the bored-out block about axis A-A. [13.54 kg.m²]
*   The mass moment of inertia of the bored-out block about axis B-B. [41.40 kg.m²]

Exercise 18.f

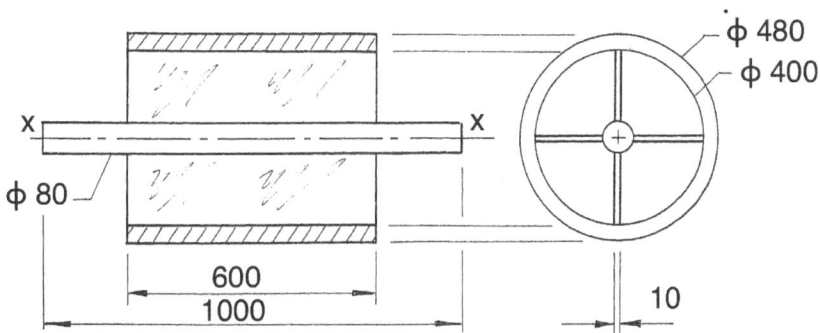

An assembly consists of a hollow copper cylinder, density 8900 kg/m³, supported by four steel plates, density 7800 kg/m³, that are welded to a steel axle, diameter 80 mm.

- Make a 3-D sketch of this assembly.
- Determine the mass moment of inertia of the assembly about axis **x-x**. [14.94 kg.m²]
- Determine the radius of gyration of the assembly [202.4 mm] and explain why this value is reasonably to be expected.

Exercise 18.g

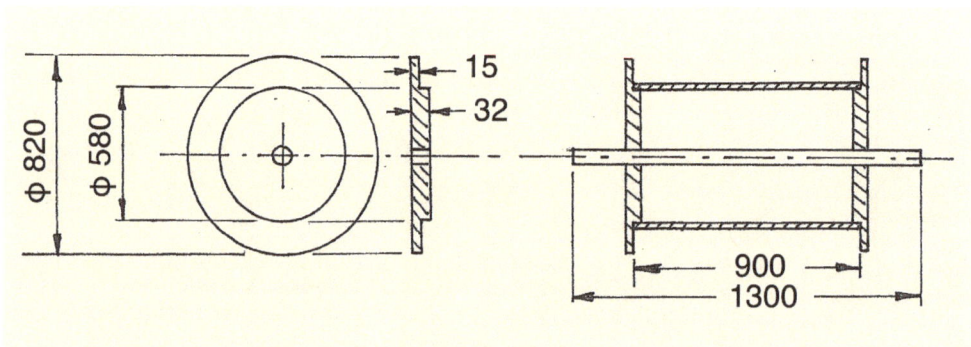

A winch drum consists of a hollow cylinder, thickness 6 mm, mounted between two stepped flanges, and a central axle, of diameter 40 mm. All parts are of steel, density 7800 kg/m³. Determine:

- An accurate value for the mass moment of inertia of this drum about its axis of rotation [22.59 kg.m²], and
- The percentage error from the value obtained above, if the contribution of the axle is ignored, and the flanges are assumed to be solid, not drilled to receive the axle. [0.0107%]

19

Rotational and linear inertia combined

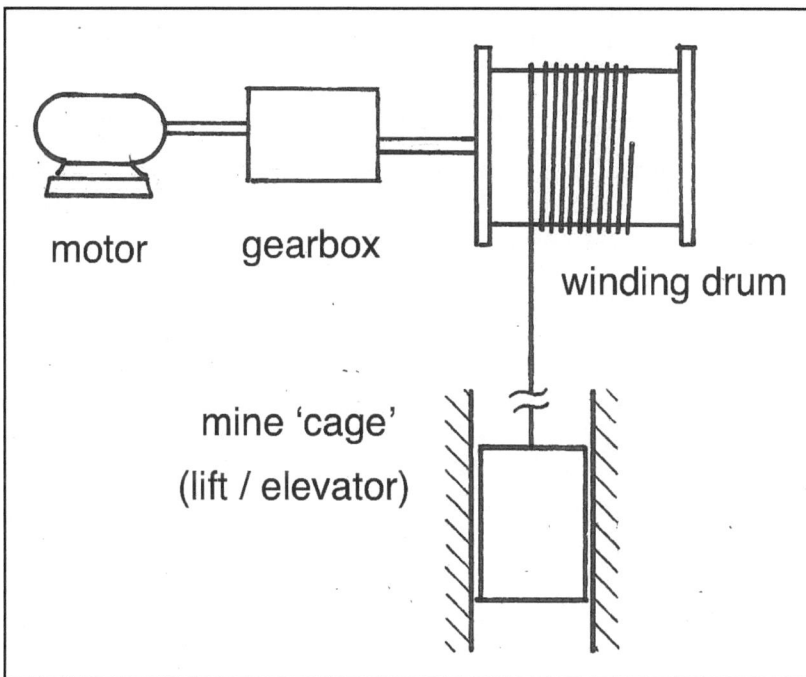

motor   gearbox

winding drum

mine 'cage'
(lift / elevator)

True/false test on this topic

	True/false Test # 19a Rotational and linear inertia combined			
1	In a system of linked rotating objects, the overall rotational acceleration has to be the same for all the objects.	T	F	?
2	In a system of linked rotating objects, the torque equation: $T = I\alpha$ has to applied separately to each rotating object to determine the angular acceleration it will experience.	T	F	?
3	A system of gears meshing in a gear train will exhibit an effective MMI that is equal to the sum of the MMI values of the individual gears in the train.	T	F	?
4	The effective rotational inertia of a system of gears meshing in the sequence **A,B,C,D** will be the same when the input gear is **A**, as it is when the input gear is **D**.	T	F	?
5	If a heavy lawn roller is caused to accelerate across a lawn in a straight line, the force required to accelerate it depends only on the linear acceleration of the mass of the roller.	T	F	?
6	Friction losses in a gear train that transmits torque to an accelerating system can diminish the torque available at the output.	T	F	?
7	Friction losses in a gear train diminish the velocity ratio of that gear train.	T	F	?
8	The torque output of an electric motor that is driving an accelerating system varies with speed.	T	F	?
9	If an electric motor has sufficient torque to drive an accelerating system, it will almost always reach its operating speed less than one second after being switched on.	T	F	?
10	The mass moment of inertia of the rotating objects in a system of linked objects is of no significance when the system is operating at constant velocity.	T	F	?

Questions requiring short descriptive answers

1. When analysing the acceleration of a system of masses linked by ropes, where the ropes pass over rotating objects such as drums or sheaves, why is it not justifiable to ignore the masses of the rotating objects?
2. Describe what is meant by the 'efficiency' of a gear train.
3. When power is transmitted by meshing gearwheels, is it possible that the velocity ratio could vary due to wear? Why, or why not?
4. Do friction losses in a gear train affect the output torque, or the output speed, or both? Explain.
5. When an electric motor drives a mechanical system such as a geared hoist, accelerating it from a standing start to a given operating speed, does the motor exert maximum power all the time? Explain.
6. Describe, with the aid of a sketch, what is meant by the 'equivalent mass' of a rotating drum that is caused to turn by a rope passing around it within a system of linked masses that move in linear paths.

Calculation exercises

Exercise 19.a

A load of 500 kg is dragged up a slope by a winch **A**. The rope passes over a drum **B** without slipping. The coefficient of friction between the load and the

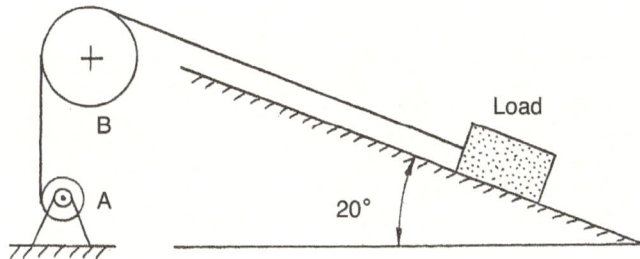

inclined plane is 0.2. The following data apply:

Winch **A**: effective drum diameter 0.4 m, mass 36 kg, radius of gyration 0.18 m, frictional torque in bearings 15 Nm.

Drum **B**: diameter 0.9 m, mass 450 kg, radius of gyration 0.42 m, frictional torque in bearings 60 Nm.

Determine the torque that needs to be applied to the winch drum to accelerate the load up the slope at 1.5 m/s². [837.9 Nm]

Exercise 19.b

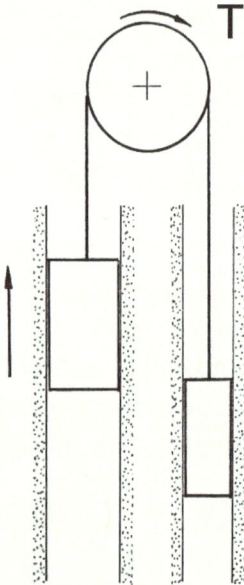

A winch raises an elevator of mass 1100 kg, assisted by a counterweight of mass 900 kg. The winch drum diameter is 720 mm, its mass is 459 kg, and its radius of gyration 340 mm. The frictional torque in the drum bearings is 45 Nm.

The wire rope makes several turns around the drum, and can be assumed not to slip.

Both the elevator and the counterweight run in guide-rails where the sliding friction force is constant at 160 N.

Determine the value of the torque, T, that must be applied to the drum to accelerate the elevator upwards at 1.0 m/s².
[1734 Nm]

If the acceleration continues for five seconds, after which time the velocity is kept constant, determine:

- the value of that velocity, [5 m/s],
- the value of the input torque during the constant velocity phase [866.5 Nm], and
- the power input needed to keep the system moving at this constant velocity. [12.04 kW]

Exercise 19.c

A rotating assembly consists of a granite wheel (density 2700 kg/m³) attached to a solid steel drum (density 7800 kg/m³) by bolts whose influence is negligible. The assembly can rotate on a fixed axle that is lubricated, and experiences a frictional torque of 6 Nm.

Determine the mass moment of inertia of the rotatable assembly.
[60.84 kg.m²]

Mass-piece **m** is attached to a light cord wound around the drum.

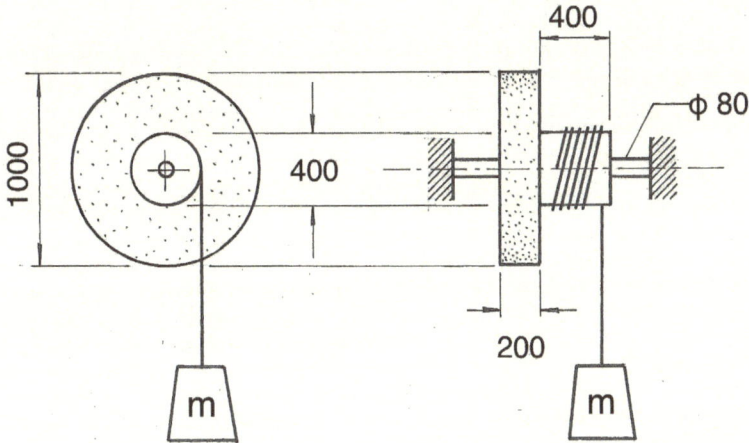

Determine the required value of **m**, if, when released, the downward acceleration of the mass-piece should be 0.1 m/s². [18.75 kg]

Exercise 19.d

Winch drum **A** is driven by an electric motor, through gearing which steps down the speed by a factor of 24. The wire rope is wound several turns around both the winch drum and the idler drum **B**, and can be assumed not to slip. The motor operating speed is 1440 r/min. The gearbox is 90% efficient.

Consider the rope diameter and mass to be negligible.

Mass-piece **C** must be accelerated upwards at 1 m/s².

Use the additional data in the table below to determine:

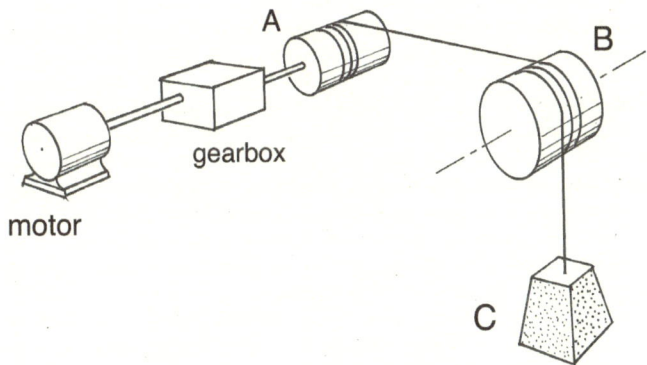

- The tension in the part of the rope between **B** and **C** [4324 N],
- The angular acceleration of the idler drum [2.5 rad/s²],
- The tension in the rope between **A** and **B** [4714 N],

- The input torque to the winch drum [987.3 Nm],
- The motor output torque required to accelerate the system [45.71 Nm],
- The time taken from a standing start until the motor reaches full speed [1.257 sec].
- The power output of the motor at the instant that it reaches full speed [6.893 kW], and
- The power output of the motor once it is operating at constant velocity [5.898 kW].

	drum A	drum B	load C
Outer diameter (mm)	400	800	n/a
Radius of gyration (mm)	180	370	n/a
Mass (kg)	90	280	400
Frictional torque in the bearings (Nm)	30	60	n/a

Exercise 19.e

A heavy lawn roller pulled by a small tractor experiences resistance to motion on account of continually needing to push its way out of the dip in the lawn surface caused by the pressure of its own weight. The resistance force for a given roller on a particular sports field is found to be 800 N.

This roller has the following specifications: diameter 720 mm, width 1.00 m, mass 1200 kg, and mass moment of inertia 78 kg.m². The frictional resistance in its bearings may be assumed negligible.

Determine the magnitude of the force that must be exerted by the tractor to accelerate this roller up a slope of 5° on the grass-covered ramp leading up to this field, with an acceleration of 0.1 m/s². [2006 N]

20

Kinetic energy of rotation and angular momentum

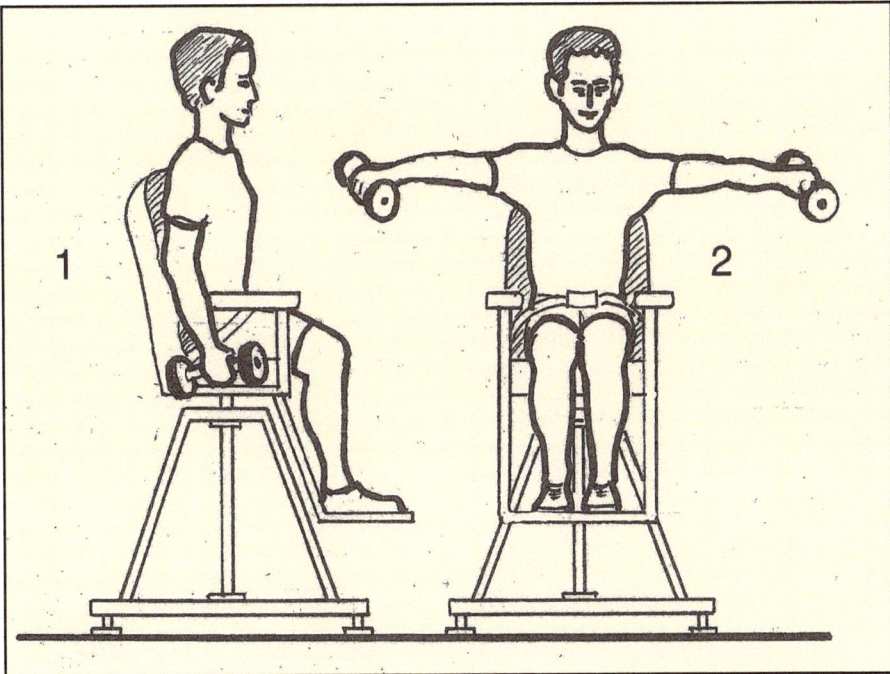

True/false tests on this topic

	True/false Test # 20a Kinetic energy of rotation, and angular momentum			
1	A rotating object possesses kinetic energy due to its capacity to perform mechanical work.	T	F	?
2	The amount of rotational kinetic energy possessed by a rotating object is directly proportional to its angular velocity.	T	F	?
3	The function of some flywheels is to store rotational kinetic energy for later use.	T	F	?
4	Rotational kinetic energy can be converted to other forms of mechanical energy.	T	F	?
5	When a wheeled vehicle accelerates, some of the driving energy has to be used to impart angular acceleration to the wheels.	T	F	?
6	The rotational kinetic energy of a rotating assembly consisting of several parts is not equal to the sum of the rotational kinetic energy values of the constituent parts.	T	F	?
7	We cannot use the principle of conservation of energy to determine how much rotational energy is transferred as a result of an angular impulse.	T	F	?
8	Angular momentum has the same units as linear momentum.	T	F	?
9	The conservation of angular momentum is the principle that makes a ballerina spin faster when extending her arms than when she had them close to her body.	T	F	?
10	The amount of angular momentum possessed by a rotating body depends on its angular velocity and its mass distribution relative to the axis about which it is rotating.	T	F	?

	True/false Test # 20b Kinetic energy of rotation, and angular momentum			
1	If a person is riding on a children's roundabout that has been brought up to speed, and that person moves towards the axis of rotation, the roundabout will slow down slightly.	T	F	?
2	An angular impulse can be caused by a torque applied to a rotatable object.	T	F	?
3	An impulsive torque could be provided by a sudden braking force on an object that is already rotating.	T	F	?
4	An object possessing linear momentum can impart angular momentum to a rotatable object.	T	F	?
5	For the same rotational speed, the angular momentum of an object rotating about an axis that does not pass through its centre of gravity would be less than that if it were rotating about a parallel axis that does pass through its centre of gravity.	T	F	?
6	The moment of the linear momentum of an object about an axis of rotation amounts to a quantity of angular momentum about that axis.	T	F	?
7	The reason why a stream-shot waterwheel has a low efficiency is that the water in the stream does not have its momentum reduced significantly on impact with the vanes.	T	F	?
8	The centre of percussion of a uniform thin rod pivoted about one end is located one third of the length of that rod away from the pivot point.	T	F	?
9	A waterwheel works by using the linear momentum of a stream of water to impart angular momentum to the wheel.	T	F	?
10	If a part breaks off a rotating object that possesses angular momentum, that part will possess a fraction of the original angular momentum.	T	F	?

Questions requiring short descriptive answers

1. What is 'kinetic energy of rotation'? Explain, and give two examples where it may be encountered.

2. If a cylinder of known mass and radius is placed at the top of a gently inclined plane, and rolls down without slipping, is it possible to determine its velocity after descending a certain distance? Why, or why not?

3. A wagon can be fitted with either of two sets of wheels, both of the same diameter. The wheels of set A have greater mass than those of set B. When fitted with set A, the wheels of set B are placed inside the wagon, and vice versa, to ensure that the total mass remains the same. If this wagon is allowed to roll from rest down a gently inclined plane from a given starting line to pass a given finish line: which journey will take longer: when fitted with set A or set B? Why?

4. Explain why so-called 'high performance wheels' make a difference to the performance of a car.

5. Define 'angular momentum'.

6. What is the function of a flywheel in a machine powered by a reciprocating piston in a cylinder?

7. Explain what the conservation of angular momentum is, and provide one example of where this phenomenon is easily observed.

8. Explain what is meant by an 'angular impulse'.

9. What is the 'effective rotational inertia' of a gear train?

10. If a number of children are standing on a rotating playground roundabout, near the circumference, and all move towards the central axis simultaneously, what effect will be noticed, and why does this happen?

11. An object with linear momentum can impart angular momentum to a rotatable object. If we know the amount of linear momentum that is available to transfer, what else do we need to know to determine the amount of angular momentum that will be gained? Explain why.

12. Describe what is meant by a 'centre of percussion' on a pivoted rod.

13. Would the position of the centre of percussion of a baseball bat be easily determined? Why, or why not?

Calculation exercises

Exercise 20.a

Six children of mass 50 kg each are riding on a roundabout in a play park. The roundabout has mass moment of inertia 1600 kg.m², and diameter 4 m. At a given moment, all the children are holding onto the radial support rails, standing at the outer edge of the roundabout, which is rotating with a peripheral speed of 3 m/s.

a. If all the children quickly and simultaneously move towards the centre of the roundabout, to a position where each of them is 500 mm from the axis of rotation, what will be the new peripheral speed of the roundabout? Ignore the effect of friction at the pivot. [5.015 m/s]
b. Describe what you think the effect of friction at the pivot would be on your answer above.
c. Describe what effect, if any, would be caused by the change to the air resistance that would occur due to the children moving to this new position.

Exercise 20.b

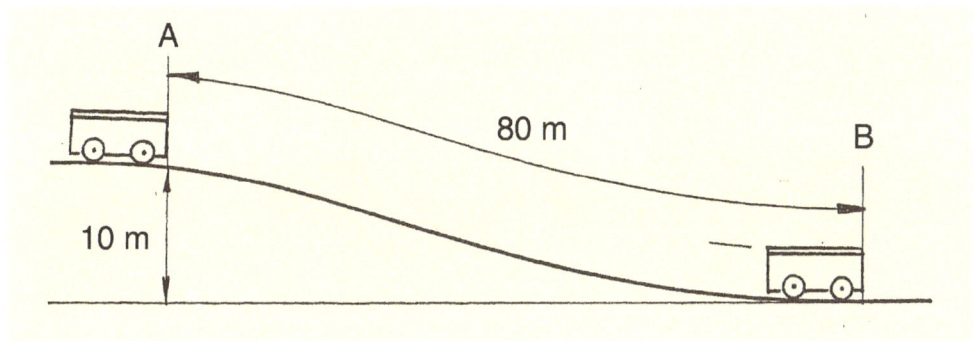

A specially constructed wagon rolls from point **A** through point **B** (a distance of 80 m along the surface), from a standing start. The mass of the wagon body is 400 kg. The frictional resistance to motion of this wagon is 180 N. The wagon is supplied with two interchangeable sets of wheels.

The four wheels of set 1 each have effective diameter 800 mm, mass 60 kg and radius of gyration 350 mm. Those of set 2 have the same diameter and radius of gyration, but are made of lighter material: their mass is 20 kg each.

Ignoring air resistance, determine the velocity of the wagon by the time it reaches point **B**, if

- The wheels of set 1 are fitted, while those of set 2 are carried in the wagon. [11.16 m/s], and
- The wheels of set 2 are fitted, while those of set 1 are carried in the wagon. [12.00 m/s]

Describe what accounts for this difference.

Exercise 20.c

A load is suspended from a light rope that is wound around a drum, to which a brake is applied. When the brake is released, the drum is free to turn, and the load descends.

The mass of the drum is 180 kg, its diameter is 1.2 m and its radius of gyration 0.5 m. The load is 40 kg.

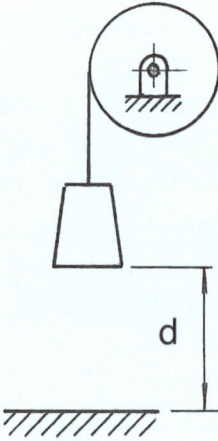

- If there were no friction in the drum bearings, how long would it take for the load to descend a distance **d** = 4 m? [1.834 sec]

- If the load descends **d** = 4m in exactly 2 seconds, determine the frictional torque in the drum bearings. [37.44 Nm]

Exercise 20.d

The rotating base of a roundabout in a children's play park has mass 800 kg, diameter 4.0 m and radius of gyration 1.5 m. It is rotating at an angular velocity of 3 rad/s while six children of average mass 50 kg are standing at the outer extreme of the rotating platform. If the children all suddenly move towards the middle and take up positions where each one's centre of mass is 0.6 m from the axis of rotation, what will be the new angular velocity? [4.717 rad/s]

Exercise 20.e

Two large gearwheels, **A** and **B**, mesh together. Consider them to be solid discs for the purposes of determining their respective rotational inertias.

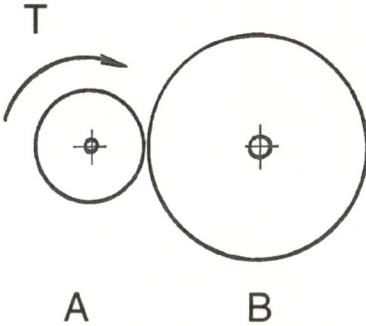

Gear **A** has diameter 400 mm and mass 20 kg. Gear **B** has diameter 800 mm and is made of the same material as gear **A**, with the same thickness.

The gearwheels are initially at rest, when a clockwise torque of 40 Nm is applied to gear **A**, and is steadily maintained. If the frictional torque on the gear shafts is negligible, determine:

- The time it will take for gear **B** to reach an angular velocity of 100 rad/s. [20 seconds] and

- The tangential force that the teeth of the gearwheels exert on one another during that time. [180 N]

- The effective rotational inertia of this gear train, from the point of view of applying a torque to gear **A**. [4.0 kg.m²]

Exercise 20.f

The back wheel of a stationary bicycle is replaced with a flywheel fixed to a sprocket whose effective radius is 25 mm. This flywheel is a solid disc of stone, density 2400 kg/m³, diameter 600 mm, and thickness 50 mm.

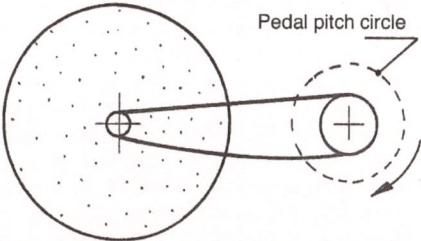

Pedal pitch circle

The rider can exert a pedalling torque of 15 N.m, and sustains this torque to bring the flywheel from rest to a speed of 3 rev/s. The front sprocket's effective radius is 75 mm. The diameter of the pedal pitch circle is 320 mm. Consider as negligible: the rotational inertia of the pedals and front sprocket, and all rotational resistance. Determine:

- The mass moment of inertia of the flywheel. [3.181 kg.m^2]
- The change in angular momentum of the flywheel during this period of acceleration. [59.96 kg.m^2/s]
- The time it takes to reach this speed. [12 seconds]
- The tension in the upper part of the drive chain. [200 N] and
- The average rate of expenditure of energy provided by the rider over this period. [47.09 W]

21

Simple harmonic motion

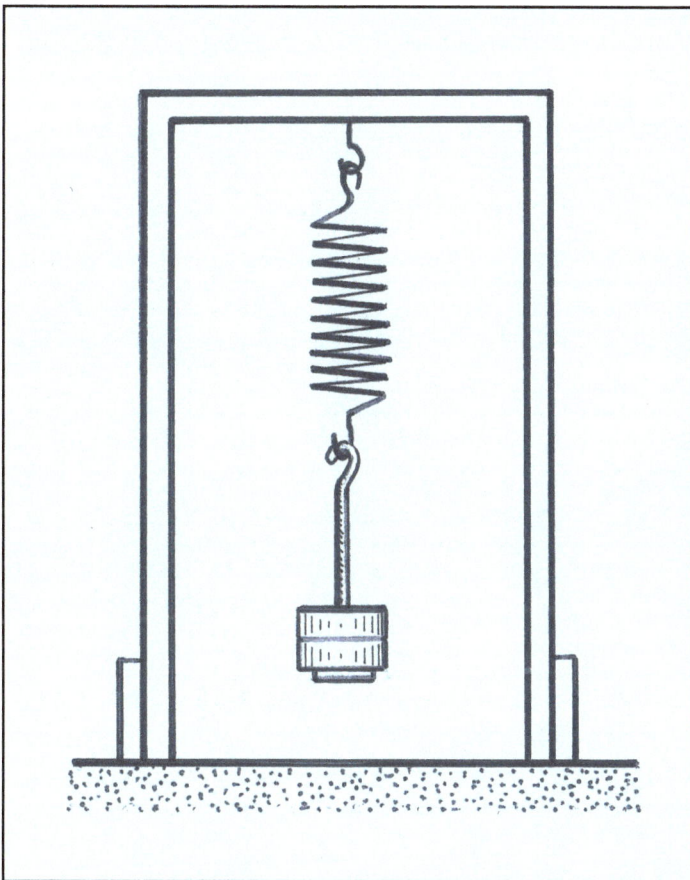

True/false tests on this topic

	True/false Test # 21a Simple harmonic motion			
1	The motion of a pendulum is an example of pure SHM.	T	F	?
2	For an object undergoing SHM, the further the object is from the midpoint of the motion, the smaller is its velocity.	T	F	?
3	The reciprocating motion of a piston inside a cylinder in an internal combustion engine roughly approximates to a SHM.	T	F	?
4	There are three conditions that define a true SHM.	T	F	?
5	The amplitude of a SHM is the distance between the two extremes of the motion.	T	F	?
6	The period of a SHM is the time taken for one complete cycle of the motion of the object undergoing the SHM.	T	F	?
7	The equation that defines the position of the object undergoing a SHM with respect to one extreme of the motion is $x = r \cos(\omega t)$.	T	F	?
8	The period of the SHM that occurs when a mass-piece suspended from a coil spring is given small vertical oscillations is dependent on the stiffness of the spring.	T	F	?
9	When a spring-mass system is left to oscillate, the amplitude of the oscillations reduces, but the period of the oscillations remains constant.	T	F	?
10	The period of a spring-mass system is the time taken for the mass-piece to move from its lowest position to when it is at its highest position.	T	F	?

	True/false Test # 21b Simple harmonic motion			
1	The period of a simple pendulum motion that approximates to SHM depends on the mass of the bob and the length of the string.	T	F	?
2	The oscillation of a compound pendulum, moving with a small initial displacement, approximates to a SHM.	T	F	?
3	The period of a compound pendulum does not depend on its mass.	T	F	?
4	Any solid object that can be suspended from a point and given small swinging oscillations could be considered a compound pendulum.	T	F	?
5	A given pendulum would have the same frequency on Mars as it had on Earth.	T	F	?
6	A given spring-mass system would always oscillate with the same frequency, no matter what the value of the gravitational acceleration was where the pendulum was placed.	T	F	?
7	Resonance occurs when an externally applied frequency matches the natural frequency of a system that can oscillate.	T	F	?
8	A circular disc suspended from a wire that coincides with its central axis, can be made to oscillate with SHM with small reciprocating rotations in the plane of the disc.	T	F	?
9	It is not possible to determine whether a given oscillatory motion is or is not a pure example of SHM.	T	F	?
10	When attempting to determine the radius of gyration of an irregularly-shaped object by allowing it to oscillate as a compound pendulum, it is preferable to suspend the object from a wire passing through a hole in the object, rather than from a knife edge.	T	F	?

Questions requiring short descriptive answers

1. Are all types of oscillatory motion 'simple harmonic'? Explain.

2. Give three examples of simple harmonic motion occurring in nature and three examples of it occurring in machinery.

3. State the three defining characteristics of a simple harmonic motion.

4. Define the amplitude of a simple harmonic motion, and indicate this on a sketch.

5. Define the period of a simple harmonic motion, and demonstrate its relation to the frequency of the motion.

6. Prove that the vertical oscillation of a mass-piece suspended from a tension spring is simple harmonic motion.

7. What is meant by the 'generating circle' of a simple harmonic motion?

8. What makes the motion of a pendulum only approximate to a simple harmonic motion?

9. Describe the difference between a simple and a compound pendulum.

10. Describe the practical use of the equation for the period of a compound pendulum.

11. What is meant by 'resonance' in relation to an oscillating system?

12. How can an object be damaged by a low-energy vibration impinging on it at its resonant frequency? Explain.

13. Describe briefly the steps in an experiment to verify the reliability of the compound pendulum equation as a means to determine the radius of gyration of a regularly shaped object.

14. For a reciprocating piston and crank system: which arrangement would more closely approximate to the piston moving with SHM: one with a small crank radius relative to the stroke of the piston, or one with a larger crank radius? Explain.

Calculation exercises

Exercise 21.a

A coil spring, of stiffness 1200 N/m, hangs vertically, suspending two mass-pieces: an upper one of 20 kg, and, hanging from it, separated by a short string, a lower one of 8 kg. The system is initially at rest.

The string between the blocks is suddenly cut, resulting in the upper mass-piece oscillating at the end of the spring. For the vibration that results, determine:

- The difference in height between the original equilibrium position of the top block and its new equilibrium position. [65.40 mm]
- The amplitude of the oscillation [65.40 mm]
- The frequency of the oscillation [1.233 Hz]
- The maximum velocity of the mass-piece during each oscillation [0.5066 m/s] and
- The velocity of the mass-piece when it is halfway between the equilibrium position and one extreme of its motion. [0.4387 m/s]

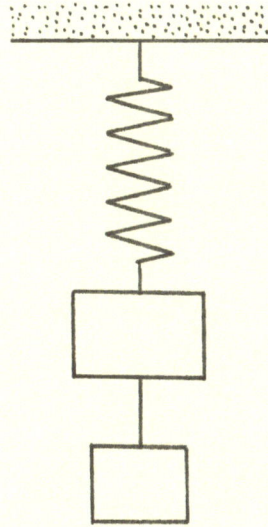

Exercise 21.b

You want to swing across a river that is 20 m wide, on a 90 m long rope that is fixed at its upper end to a point on a bridge above the centre of the river. If you don't push off from the bank, but simply raise your feet from the ground, how long will it take you to reach the other side? [9.5 sec]

Exercise 21.c

In order to select two appropriate matching coil springs for a swing-seat of the type shown below:

Suppose that two people of mass 80 kg sit down gently on the 16 kg swing-seat, causing the springs to extend by 100 mm from the no-load position to an equilibrium position.

Determine the required value of the stiffness of each spring. [7848 N/m]

If the people getting onto the seat were to sit down suddenly (a case of sudden loading), how far below the no-load position would the springs extend? [200 mm]

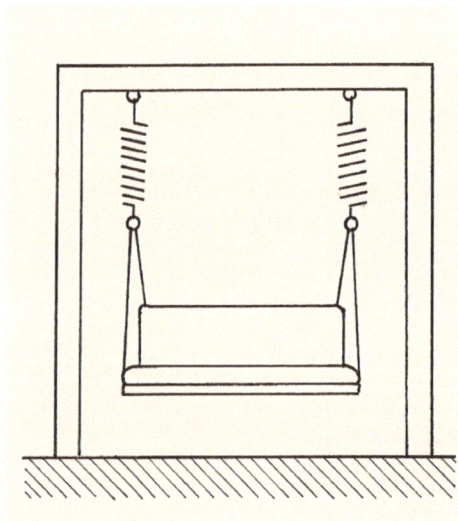

If a coil spring should not be required to extend more than 30% of its initial length, determine a suitable value for the un-stretched length of each spring. [667 mm]

Determine the time it would take for 5 complete oscillations if the people got onto the seat suddenly, and the seat bounced until it eventually came to rest in the equilibrium position. [3.327 seconds]

Exercise 21.d

a. In a place where 'g' has the value 10 m/s², the pendulum of a clock has a period exactly 2 seconds. If someone lengthened the pendulum by 1 mm, how many seconds would it gain or lose in a day of 24 hours? [loses 21.31 seconds]

b. On planet X , if you swing a pendulum that is 800 mm long, it is found to take 23 seconds for 10 complete oscillations. What is the value of 'g' on that planet? [5.970 m/s²]

Exercise 21.e

A coil spring is suspended from an overhead frame. Initially, it supports a weight hanger of 500 g and a mass-piece of 5 kg. When a second mass-piece of 5 kg is added, the spring stretches as shown.

From this position, the load hanger is pulled downwards a further 60 mm. It is then is released, allowing the system to oscillate.

- Sketch a graph of force vs. extension for this spring, to scale.
- Determine the stiffness of this spring. [352.9 N/m]
- Determine the period of the resulting oscillations. [1.084 sec]
- During the oscillation, what will be the velocity of the mass-pieces when they are 10 mm from the extremes of their range of motion? Ignore the diminishing amplitude due to energy loss. [0.192 m/s]

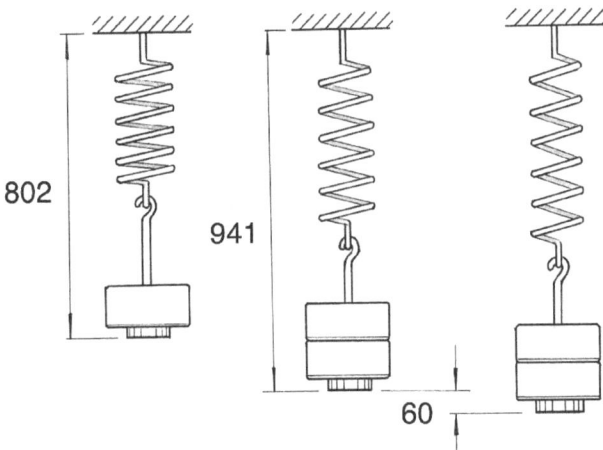

Exercise 21.f

This cast-iron flywheel has a complex shape, making it difficult to establish its mass moment of inertia by directly calculating the way the mass is distributed.

However, the value of the MMI can be determined by allowing it to swing with small oscillations in the plane of the wheel, suspended from a knife edge as shown, and timing the oscillations.

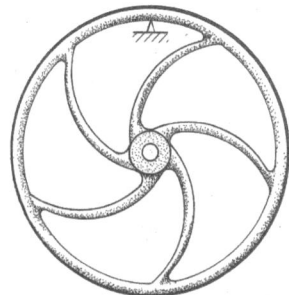

If the mass of the flywheel is 242 kg, the knife edge is 465 mm from the centre of the wheel, and the time taken for 10 oscillations is 19.24 seconds, determine the MMI of this wheel about its central rotation axis. [51.18 kg.m²]

Exercise 21.g

There are two pendulums:

The first is a simple light string, of length 'L', with a small spherical mass-piece of mass 960 g attached at its lower end. The other is a slender rod, of length 1000 mm and mass 960 g, which is suspended from a pivot point that is 100 mm from its upper end.

If these pendulums have to oscillate with the same frequency, determine the dimension 'L'. [608.3 mm]

Exercise 21.h

An irregularly-shaped object, made from plate of uniform thickness and density, is suspended so that it can swing freely about a pin through point **O**. The pin is aligned on axis **A - A** that is perpendicular to the plane of the plate. Point **O** is situated 376 mm from point **G**, the centre of mass of the object.

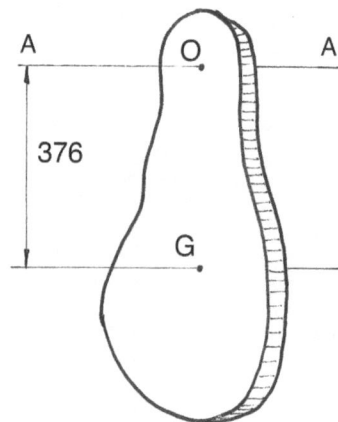

The mass of the object is 24.84 kg. When caused to swing with small oscillations, it is found to complete 8 beats in 6.6 seconds.

Determine the mass moment of inertia of the object about an axis parallel to **A - A**, that passes through point **G** [2.807 kg.m²] and its mass moment of inertia about axis **A - A** [6.319 kg.m²]

22

Basic
vehicle dynamics

True/false tests on this topic

	True/false Test # 22a Basic vehicle dynamics			
1	A self-propelled vehicle moves forward due to the horizontal component of the ground reaction force on the driving wheels.	T	F	?
2	The road reaction force acting on a wheel is identical, whether the wheel is passive or is driven by a power unit on the vehicle.	T	F	?
3	The ground reaction force on a wheel is an external force that should be included in a free-body diagram of a vehicle.	T	F	?
4	The towing force needed to accelerate a trailer with a given value of acceleration depends only on the mass of the trailer.	T	F	?
5	The torque needed to turn a driving wheel on a self-propelled vehicle when moving at constant velocity is given by $T = D.r$ where D is the horizontal component of the ground reaction force acting on that wheel and r the effective radius of the wheel.	T	F	?
6	The amount of torque produced by any engine is not constant at all speeds.	T	F	?
7	An internal combustion engine that drives a vehicle can be presumed to be functioning at its rated power under all conditions.	T	F	?
8	The amount of power being put out by an engine can only be determined if we know the rate at which it is actually performing work.	T	F	?
9	One horsepower (HP) is roughly 1.5 kW.	T	F	?
10	The work-rate which James Watt called one horsepower was based on the maximum work-rate of all the horses he tested.	T	F	?

	True/false Test # 22b Basic vehicle dynamics			
1	The overall transmission ratio of a self-propelled vehicle is the ratio between the speed of the input shaft to the gearbox and the speed of the output shaft of the gearbox.	T	F	?
2	The diameter of the driven road-wheels is a factor in determining the traction that these wheels can achieve.	T	F	?
3	In first gear, the engine speed is much greater than the road-wheel speed.	T	F	?
4	In top gear, the overall gear ratio of a vehicle can sometimes reach values of 6 or more.	T	F	?
5	The transmission efficiency is related to the drop in speed of the output shaft, compared with the speed that it ought to have if it were 100% efficient.	T	F	?
6	Friction in the drive train of a vehicle reduces the value of the torque reaching the driven wheels.	T	F	?
7	A self-propelled vehicle will experience wheel-spin if the limiting friction force where the drive wheels are in contact with the road has been reached.	T	F	?
8	The reason for tyres heating up in use is that they experience sliding friction with the road.	T	F	?
9	The magnitude of the air resistance force (known as drag) experienced by a vehicle depends on only two variables.	T	F	?
10	The frontal surface area of a vehicle is one of the factors that determines the amount of drag experienced by the vehicle.	T	F	?

Questions requiring short descriptive answers

1. What is the essential difference between the force exerted by the road on (a) the wheels of an externally driven vehicle such as a trailer, and (b) the driven wheels of a self-propelled vehicle, such as a car?
2. What external force propels a self-propelled vehicle forwards?
3. List four ways in which towing a trailer can place demands on the towing vehicle.
4. When analysing the forces on a wheel that is moving with constant velocity, do we need to take into account the rotational inertia of the wheel? Why, or why not?
5. Define 'the overall transmission ratio' of a car.
6. Explain why output torque (but not output speed) is affected by the efficiency of a gear train. What causes this effect?
7. Under what conditions would wheel-spin occur when a car is being accelerated?
8. Do car tyres heat up as a result of friction with the road surface, or from some other cause? Explain.
9. Describe how to measure the rolling resistance of a motor vehicle.
10. What is the main factor that determines the maximum allowable speed of any given internal combustion engine?
11. Explain what the 'drag coefficient' is.
12. Does the air resistance on a vehicle constitute a significant force at highway speeds? Explain.
13. What is the purpose of 'streamlining' a vehicle?
14. If an engine is rated as developing a certain value of torque, does it actually develop this value at all speeds? Explain with the aid of a sketch graph.
15. Name the factors that diminish the amount of torque that is available to turn the drive wheels of a vehicle driven by an internal combustion engine.
16. Would two vehicles of exactly the same shape, of but different sizes, experience the same amount of air resistance at the same speed? Explain.
17. Would you classify a bicycle as a self-propelled or an externally driven vehicle? Explain.

Calculation exercises

Exercise 22.a

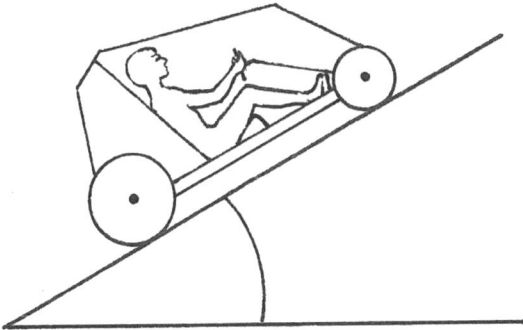

A car built for a student competition has mass 200 kg and carries a driver of mass 80 kg. Its engine puts out 7.5 kW at 2500 r/min. The rolling resistance is 45 N. The diameter of the driving wheels is 500 mm. Assuming a gear ratio of 15:1 and a transmission efficiency of 78%:

- What is the angle of the steepest slope this car could climb without losing speed? [28.15°]
- What would this angle be, if the overall mass were reduced by 30 kg? [31.90°]
- What would this angle be, if, in addition to the mass reduction, the driving wheels were exchanged for ones with diameter 400 mm? [41.69°]

Exercise 22.b

If you need to determine the average power output during the period for which a vehicle drives up an incline while accelerating, over a given distance: which of the following values do you need to know: (circle one option for each item listed). You have to get all of them correct, or your score for this question will be zero. 'yes' = this info is needed, 'no' = this info is not needed.

Mass of the vehicle	yes	no
Efficiency of the transmission system	yes	no
Coefficient of friction between the tyres and the road	yes	no
Value of the gradient	yes	no
Value of the rolling resistance	yes	no
Maximum power rating of the engine	yes	no
Transmission ratios to be used	yes	no
Initial velocity of the vehicle	yes	no
Final velocity of the vehicle	yes	no
An indication of the air resistance to be expected, as a function of speed	yes	no

Exercise 22.c

A motor vehicle of mass 1200 kg experiences rolling resistance of 250 N, assumed constant, irrespective of speed.

- Determine the magnitude of the driving force at the wheel circumference, required to drive this vehicle up a slope of 20° at a constant speed of 72 km/h. Consider the air resistance on this vehicle to be 300 N at this speed. [4576 N]

- What needs to be the power output of the engine to accomplish this? [91.53 kW]

- If this same vehicle is accelerating up this slope at 1 m/s^2, at the instant when its speed is 72 km/h, what instantaneous power output is required? [115.5 kW]

Exercise 22.d

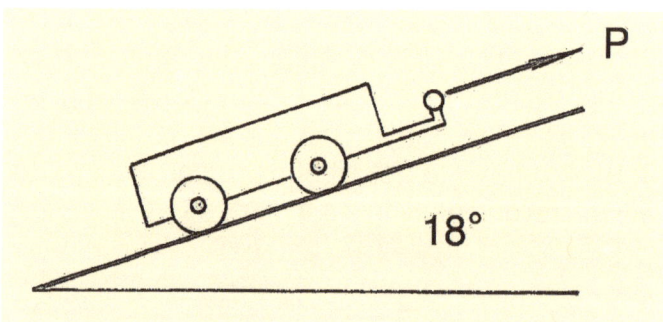

A large trailer wiith total mass 2080 kg, and rolling resistance assumed constant at 320 N, is pulled up an inclined road by a truck.

Each of its four wheels has mass 44 kg, effective diameter 800 mm and radius of gyration 360 mm.

Ignore the effect of air resistance, and determine the value of the towing force, **P**, required to pull this trailer up the slope:

- With an acceleration of 1.2 m/s² [9293 N], and
- At constant velocity. [6625 N]

Also determine the power drain on the truck due to pulling this trailer up this slope, when the speed is 90 km/h, while accelerating at 0.2 m/s². [176.7 kW]

Exercise 22.e

A 400 tonne train on a horizontal track approaches point **A** at the bottom of a long straight incline of 1 in 70, doing 27 km/h. From point **A**, the train begins to accelerate with uniform acceleration, until it reaches point **B**, which is 3 km away, at the top of the incline. By the time it reaches point **B**, it is moving at 81 km/h.

The rolling resistance of this train is 64 kN. The air resistance on this train is 1 kN at 27 km/h and 9 kN at 81 km/h.

Make a neat sketch, showing all the relevant forces acting on the train while ascending this incline. Determine:

- The value of the acceleration [0.075 m/s²],
- The increase in the train's gravitational potential energy from point **A** to point **B** [168.2 MJ],
- The work done against the rolling resistance from **A** to **B** [192 MJ],
- The driving force at the wheel circumference necessary to accomplish this motion at both point **A** [151.1 kN] and at point **B** [159.1 kN], and
- The power output required of the engine at point **A** [1.133 MW] and at point **B** [3.579 MW]

If the air resistance (drag force, F_d) experienced by this train is a function of the square of the velocity, determine what that function is, and the amount of work done against air resistance while travelling between **A** and **B**. [F_d = 17.778 v^2; 65 kJ]

Exercise 22.f

A certain sedan car has the following specifications:

- Mass of car plus driver = 1600 kg,
- Effective wheel diameter = 600 mm,
- Overall transmission efficiency = 86%
- Air resistance (drag force, F_d) on this car is dependent on velocity according to the equation: $F_d = 0.45\ v^2$,
- Rolling resistance of the vehicle is 50 N at all speeds,
- Engine output torque (assume constant at all speeds, although torque varies with speed in reality) = 300 Nm
- Overall transmission ratios in the four forward gears are: first gear n = 6, second gear n = 4, third gear n = 2, top gear n = 1,
- Maximum allowable engine speed = 5000 r/min, to avoid engine damage.

Determine:
- The maximum possible speed of this car up a slope of 10° in first gear [28.27 km/h]
- The maximum possible speed of this car up this slope in any gear, and which gear that is [138.3 km/h in second gear, although the engine speed would be approaching maximum revs, and it would be sensible to slow down]

Explain in words why it was not necessary in this exercise to provide data that would enable mass moment of inertia of the wheels to be established.

23

A variety of topics in basic mechanics

The exercises in this set are suitable for tackling in the later stages of a course, when students have been exposed to a number of topics and are expected to be able to answer questions on any of the topics that they have studied.

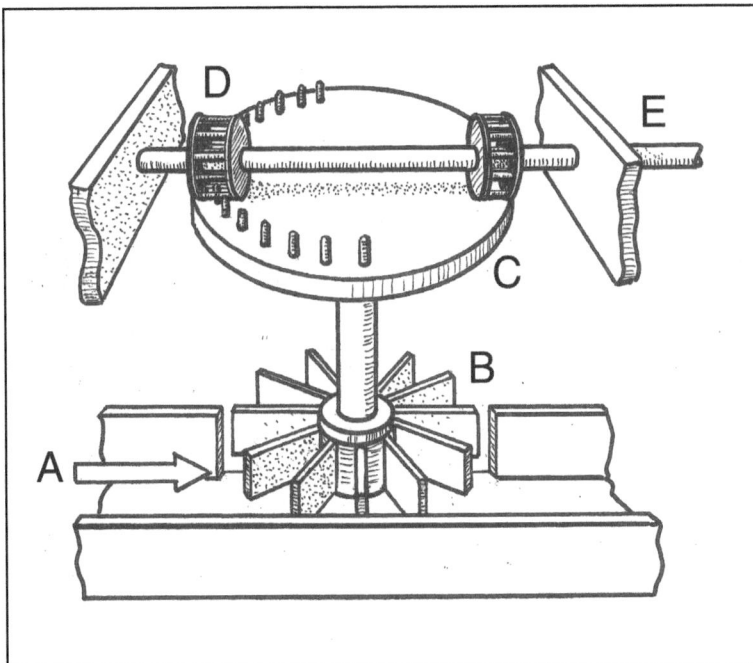

True/false tests

	True/false Test # 23a On a variety of topics in mechanics			
1	A force couple acting on an object has a different effect to that of a torque applied to the object.	T	F	?
2	What led to the S.I. system of units being developed was frustration among scientists that units in use in different countries were not equivalent.	T	F	?
3	The graphical construction known as a polygon of forces can be used to determine a maximum of three unknowns in a set of co-planar forces.	T	F	?
4	A distributed load on a beam is equivalent to a single force of the same value as the weight of that load, acting through the centre of gravity of that load.	T	F	?
5	There is always the same amount of mass on either side of the centre of mass of a rigid body, from whichever angle it is viewed.	T	F	?
6	If a long steel member of a pin-jointed truss has to be in compression, it would be a good idea to make it consist of two members joined at a node, and insert no-load members that join this node to others.	T	F	?
7	A pincer grab closes on the object being raised because the friction force between the jaws and the load contributes to the jaws increasing their pressure on the object.	T	F	?
8	The trajectory of a projectile can be analysed if the launch velocity is known, if air resistance is ignored.	T	F	?
9	On a graph that illustrates the Law of a simple lifting machine, it is possible for the graph to intercept the load axis.	T	F	?
10	When a fluid stream has its velocity changed by interacting with a paddle wheel, for us to determine the force it exerts on the paddle wheel, we need only know the velocity of the fluid stream before and after the collision.	T	F	?

	True/false Test # 23b On a variety of topics in mechanics			
1	The drag force on vehicle can be ignored at very low speeds.	T	F	?
2	In a system of linked rotating objects, the overall rotational acceleration has to be the same for all the objects.	T	F	?
3	Air resistance plays a minimal role in affecting the trajectory of a high speed projectile.	T	F	?
4	If a truss has a rigid member that passes through one or more nodes, the forces in the members cannot be analysed by the method of considering the equilibrium of forces.	T	F	?
5	When a vehicle is said to accelerate an occupant with a force of five 'g's, that means the force between the occupant and the seat back is equivalent to five times his or her weight.	T	F	?
6	If the mechanical advantage of a simple lifting machine is high, the distance moved by the effort must be considerably more than the distance moved by the load.	T	F	?
7	If a pair of pliers is being used to exert a known force, one can determine the shear force on its pivot pin by making use of free-body diagrams of all the parts of the pliers.	T	F	?
8	The friction loss in a pulley block that is part of a lifting arrangement, is entirely due to the rope sliding relative to the pulleys.	T	F	?
9	A 'particle', as conceived in mechanics, is a convenient fiction.	T	F	?
10	Air resistance is the main factor that determines the upper limit of the speed of a land vehicle.	T	F	?

	True/false Test # 23c On a variety of topics in mechanics			
1	The value of the gravitational acceleration 'g' is constant at all points on the Earth.	T	F	?
2	In a block and tackle, the diameter of the sheaves does not make a difference to the efficiency of the lifting arrangement.	T	F	?
3	The weight of a pin-jointed truss could be significant in determining the forces carried by the individual members when under load.	T	F	?
4	If it wasn't for air resistance, a projectile fired at a given speed over horizontal ground would have the same range when the launch angle is 23° as when it is 67°.	T	F	?
5	The location of the centre of gravity of an object is in all cases identical with that of its centre of mass.	T	F	?
6	It is impossible to determine the launch speed of a projectile such as an arrow, by purely mechanical means.	T	F	?
7	Rocket engines work on the principle of Conservation of Momentum.	T	F	?
8	All categories of mechanical brake operate using friction.	T	F	?
9	A system of gears meshing in a gear train will exhibit an effective MMI that is equal to the sum of the MMI values of the individual gears in the train.	T	F	?
10	To analyse the motion of a self-propelled vehicle, requires a free-body diagram of the vehicle and the simultaneous application of the torque equation.	T	F	?

	True/false Test # 23d On a variety of topics in mechanics	T	F	?
1	The effective mass moment of inertia of a system of meshing gears arranged in the sequence **A,B,C,D** will be the same when the input gear is **A**, as it is when the input gear is **D**.	T	F	?
2	There is no difference between the precision and the accuracy of a set of readings.	T	F	?
3	If there is a riveted plate at any node in a truss, that plate could transmit a force moment from one member to others, rendering the solution of the structure indeterminate.	T	F	?
4	A human-powered water-craft could be made to move by taking in water at the front of the craft and pumping it backwards to emerge at the back of the craft.	T	F	?
5	The effect of air resistance on a projectile is more noticeable in the vertical direction than in the horizontal direction.	T	F	?
6	Friction is the only cause of the efficiency of a simple lifting machine being less than 100%.	T	F	?
7	At the top speed of most production cars, the drag force would be approximatey 100 N.	T	F	?
8	In a system of linked rotating objects, the torque equation: $T = I\alpha$ has to applied separately to each rotating object to determine the angular acceleration it will experience.	T	F	?
9	The most reliable way of determining the mass of an object is to accelerate the object using a known force and to measure the resulting acceleration.	T	F	?
10	The principles of Conservation of Momentum and Conservation of Energy contradict one another.	T	F	?

Calculation exercises

Exercise 23.a

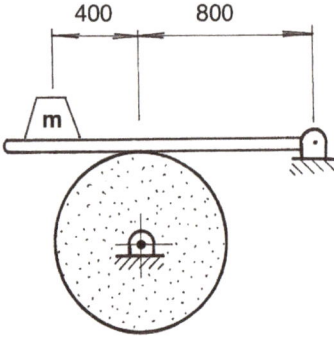

A grinding wheel made of stone, density 2500 kg/m³, thickness 120 mm and diameter 800 mm is mounted on bearings in which the frictional torque is 4.0 Nm. A light lever carrying a point mass **m** can be lowered onto the wheel.

The wheel is brought up to a speed of 90 r/min, at which speed the driving torque is removed, and at the same moment the lever is lowered onto the wheel, bringing it to rest with uniform deceleration in 10 seconds. The coefficient of friction between the lever and the wheel is 0.5.

Determine:

*   The mass moment of inertia of the wheel. (Ignore the central hole and the axle that passes through this hole: consider it as a solid disc.) [12.06 kg.m²]
*   The value of the uniform deceleration. [0.9425 rad/s²]
*   The magnitude of mass **m**. [2.504 kg]

Exercise 23.b

A coil spring is suspended from a fixed point, with a weight hanger of mass 500 g hanging from it. Without any added mass-pieces, dimension **h** = 659 mm.

Mass-pieces are now added progressively to the hanger, and the values of h are recorded each time.

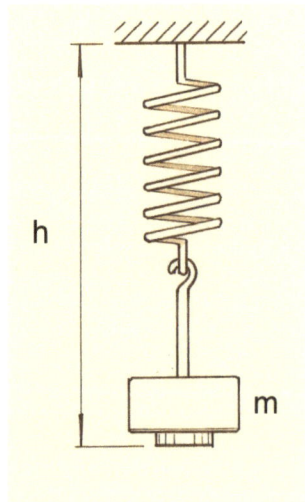

when **m** = 2 kg, **h** = 723 mm
when **m** = 5 kg, **h** = 817 mm
when **m** = 7 kg, **h** = 880 mm
when **m** = 10 kg, **h** = 976 mm

- On graph paper, draw a graph to determine the value of the spring constant, or stiffness. [310 N/m]
- What value of mass, **m**, should be placed on the hanger to result in oscillations with a period of exactly one second? [7.352 kg]
- Would a pair of springs with the same spring constant as this one be suitable for powering a spear-gun? Assume they will need to be capable of a maximum extension of 600 mm. Explain in words, with a justifying calculation: why, or why not.

Exercise 23.c

An unmanned wagon freewheels down a hill from rest at point **A**, through

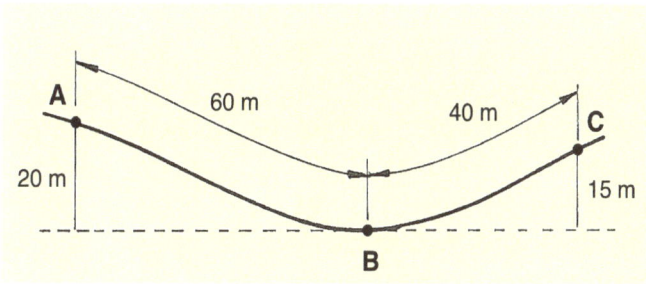

point **B** and up the further slope, coming to rest at point **C** momentarily, before rolling down again.

The total mass of the wagon is 220 kg. It has four wheels with rubber tyres. The mass of each wheel is 12 kg, its effective diameter is 680 mm, and its radius of gyration 270 mm. Assume that, at the relatively slow speeds achieved, air resistance may be ignored.

Determine:

- The rolling resistance of the wagon. [107.9 N]
- Its velocity when it goes through point B for the first time. [17.12 m/s]
- The vertical height that it rises up slope BA after passing point B, once it has rolled back from point C. [approximately 11 m]

Exercise 23.d

An acrobat whose mass is 80 kg holds onto the crossbar of pair of parallel ropes that are 9 m long, and have been pulled to one side of a vertical position by 4 m. If he casts off from his platform by raising his feet, without jumping, determine:

- The maximum tension in each rope at any time during the swing, to the nearest newton. [474 N]

- Assuming the swing occurs with simple harmonic motion, how long will it take him to swing back to his original position? (to the nearest second) [6 seconds]

If the initial sideways displacement was 3 m instead of 4 m, how would that affect the above two answers? Do not calculate, just describe your reasoning.

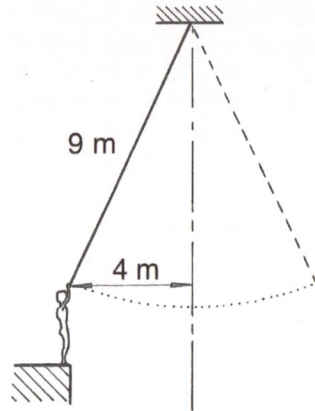

Exercise 23.e

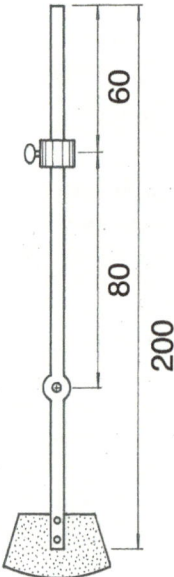

Shown here is the essential structure of the compound pendulum of a mechanical metronome that provides beats at a constant tempo for musicians. Not shown is the escapement mechanism and spring which provides the pendulum with a small input of energy on each beat to counteract die-off.

If the mass of the rod (of uniform section) is 100 grams, the fixed mass is 300 grams (consider it as a point mass), and the mass of the adjustable weight is 120 grams, what tempo would this metronome provide, in beats per minute? Consider the adjustable weight also to be a point mass. [94.44 beats per minute]

Exercise 23.f

A car of mass 1432 kg goes over a rise in a country road, where the vertical radius is **r**. The mass of the driver is 84 kg.

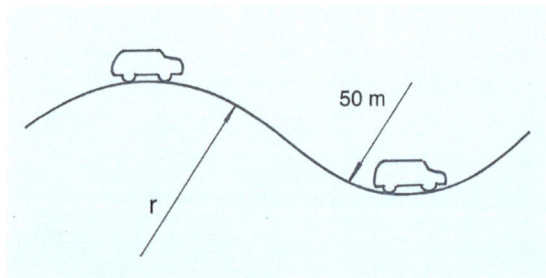

Determine:

- The minimum allowable value of r, if the car is to maintain a speed

of 72 km/h without leaving the road at the top of the rise. [40.77 m]
- The normal force between the car and the road at the top of the rise if the velocity is 72 km/h and the value of r is 60 m. [4765 N]
- The number of g's experienced by the driver at the bottom of a dip in the road with vertical radius 50 m at the same speed. [1.815]

Exercise 23.g

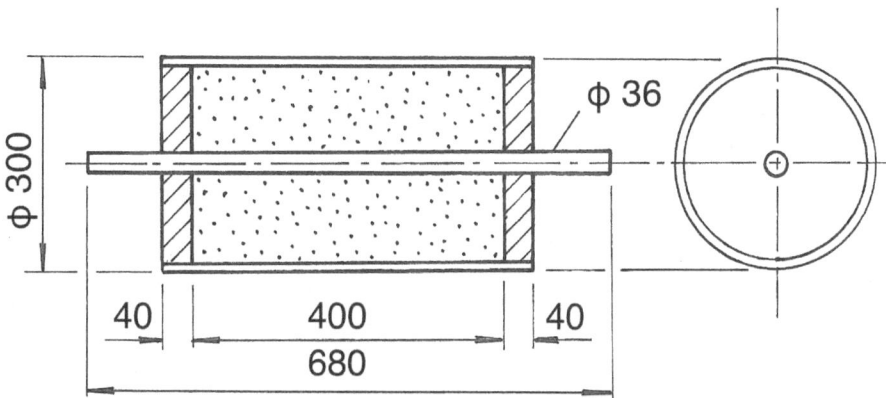

A roller drum consists of a steel axle, two flanges made of copper, and a thin-walled cylinder made of copper, 10 mm thick.The space inside the cylinder is filled with concrete, of density 2300 kg/m³. Take the density of steel to be 7800 kg/m³ and that of copper to be 8900 kg/m³.
Determine:

- The mass of the assembly [143.1 kg] and

The mass moment of inertias of:

- the axle [0.8746×10^{-3} kg.m²] ,
- one flange [0.2148 kg.m²] ,
- the thin-walled cylinder [0.8183 kg.m²] ,
- the whole assembly [1.8037 kg.m²] ,
- The radius of gyration of the assembly [112.25 mm] ,
- The braking torque required to bring the cylinder to a stop in 3 seconds, from a rotational speed of 2 rev/second [7.555 Nm] and
- The magnitude of the centrifugal force on the bearings of this cylinder when it is rotating at 2 rev/second, if its centre of gravity is displaced from its axis of rotation by 1 mm. [22.61 N]

167

24

Exercises suitable for tackling in groups

Note:

1. *These exercises can be tackled by individuals, if desired. However, the discussion that takes place in a small group can considerably benefit the process of finding out how to approach a solution. A small group would ideally consist of 3 members. Four is acceptable, but with five or more there will always be some participants who don't get a chance to contribute.*
2. *For many of these exercises, the answers depend on the line of reasoning taken by the participants, hence numerical values cannot always be supplied.*
3. *Some of the questions do not have numerical answers. This is quite normal for many situations encountered in engineering. Sometimes it is sufficient to have a knowledge of the principles of mechanics, so you can point your thinking in an appropriate direction for the next decision. It's the logic that counts.*

Exercise 24.1 Rolling resistance apparatus

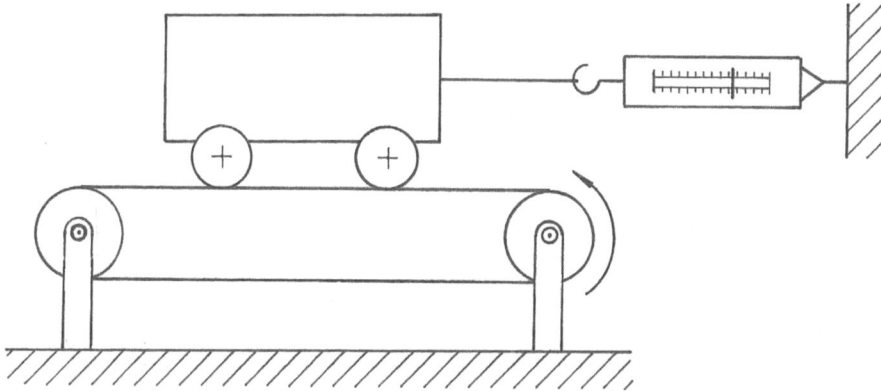

This schematic diagram shows an apparatus proposed to determine whether the rolling resistance of a given small wagon varies with speed and/or load. The speed of the conveyor belt 'treadmill' can be varied and measured. Additional masses can be placed in the wagon. All four wheels on the wagon are free to rotate.

Identify which of the following factors could potentially affect the results, and which ones may be ignored for the purposes of this proposed experiment:

- The tension of the treadmill belt
- The coefficient of friction between the tyres and the belt
- The diameter of the wheels
- The mass of the wagon
- The frictional resistance of the wheel bearings
- The alignment of the spring scale
- The accuracy of the spring scale
- The precision of the spring scale
- The duration of the test
- The ambient temperature

Give your reasons for all your selections.

Exercise 24.2   Gravity-powered vehicle

Suppose you are required to design and build a wheeled vehicle that will travel as far as possible along a smooth, flat floor, in a straight line, using the energy given up by a descending mass-piece.

The cast-iron mass-piece you may use would be either a 1 kg piece descending through a height of 1 m, or a 2 kg piece descending through 0.5 m.

The allowable materials are wood, hardboard, string, glue, rubber and lubricant. The mass-piece must remain on the vehicle, and the vehicle must have at least three wheels in contact with the floor, for the whole of its travel. There is no limit to the mass of the vehicle. You may choose the type of string you use, but may only have one type of string on your vehicle.

Attempt to answer the following questions in your group:

1. For this purpose, what are the advantages and disadvantages respectively of the two sizes of mass-piece?
2. How would you transfer the energy from the descending mass-piece to the wheels?
3. What size wheels would be best? As large as possible, as small as possible, or would there be some optimum size?
4. Would you consider using variable gearing? If so, how would you achieve a variable gear ratio with the materials at your disposal?
5. What kind of string would you think is most suitable? Why?
6. What are the advantages and disadvantages of having rubber tyres on your vehicle?
7. What effect will the mass of the vehicle have on its performance?
8. Which features would you be trying to minimise?
9. Can you think of other questions to ask that would affect your approach to the work?
10. Which mechanical principles, laws and concepts seem to apply here?

Exercise 24.3   Ritchel's dirigible

In 1878, Prof Charles F. Ritchel designed and built a dirigible which flew at an exhibition in Philadelphia, piloted at first for an indoor flight by a young woman, Mabel Harrington, and subsequently outdoors by a teenage boy, Mark Quinlan. Ritchel couldn't operate it himself, as a grown man's weight would have been too much for the amount of buoyancy provided by the hydrogen-filled blimp.

The illustration here (from the cover of Harper's Weekly of July 13, 1878) shows his 'flying bicycle' in flight. The artist fancifully put Ritchel at the controls, typical of media distortion.

Some details: the blimp was a cylinder of fine linen with a rubber coating, containing hydrogen. It was 25 ft long and 13 ft in diameter, and weighed 66 pounds.

The frame was made of hollow brass tubing. The total weight of the machine was 112 lb. The pilot weighed 96 lb.

Power for the propeller was provided by a hand crank, similar to a bicycle pedal crank, but with a larger sprocket. In order to provide directional control, the propeller axis could be turned by a foot-operated control mechanism. The first outdoor flight in Hartford, Connecticut, reportedly went out over the Connecticut River and back, landing at the starting point.

Some questions:

- Check on the buoyancy to be expected from this blimp. Would it have been sufficient to keep the machine and pilot airborne?
- Do you think the members connecting the blimp to the lower frame were solid or flexible? Why? Which option would have been preferable?
- Nowadays, what material could we use in place of the rubberised linen? Substantiate your suggestion.
- Would it be advantageous to replace the brass tubing with another material? Which?
- How much of a weight saving could be achieved with the materials you suggest?
- How do you think the mechanism that steered the propeller worked?
- How much power could a teenager exert in this fashion?
- What sort of speed do you think could be achieved using this means of propulsion? Can this speed be estimated by a plausible calculation?
- How would you control the altitude of such a craft?
- Would a different shape of blimp incur less air resistance? What shape would you suggest?
- There are no safety features evident in this drawing. If you were to build one today, what safety features would you incorporate? *Apparently on a subsequent flight, at an altitude of about 200 ft, the propeller gearing jammed, so the craft could not be steered. The boy pilot inched forward along the gantry, tied one ankle and his left wrist to the frame, then, hanging from the frame, he unjammed the gearing with his pocket knife, before working his way back to the seat and continuing.*

Exercise 24.4   Human-powered water craft

Humans have made rafts and boats move on water for millennia. There are a number of ways in which the force exerted by a person on a floating object can be made to achieve thrust. Traditional modes of providing thrust are paddles and oars. More recent developments include pedal-powered submerged propellers, paddle-wheels, and air propellers.

Illustrated here is an alternative design that makes use of a simple flap valve as found in a lift pump. There are two cylinders, one inside the other. The outer one is fixed to the craft. The operative idea is that, on the pull stroke, the flap valve closes, so the water in the inner cylinder is pushed backwards, providing thrust. On the forward stroke, more water enters the inner cylinder.

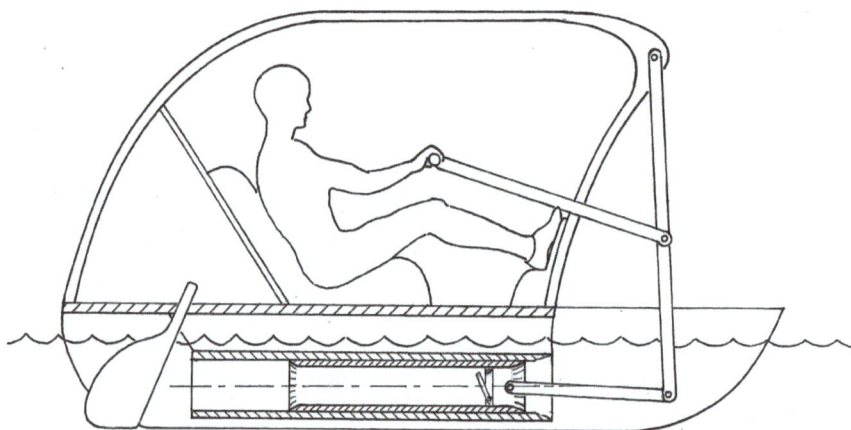

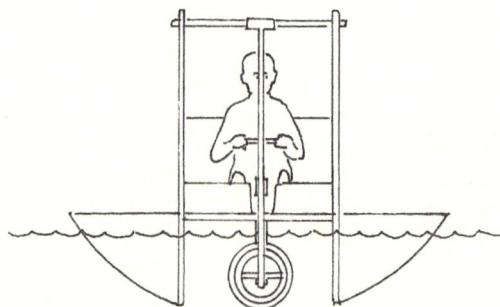

The craft is double-hulled. A rudder is shown, but the linkage that controls it is not shown, as the means of steering is not relevant to the questions about propulsion below:

1. Would this concept for achieving thrust actually work?
2. If it would work, which principle or law of mechanics explains *why* it works?

3. On the forward stroke, is there any resistance offered by the water? If so, where? What could be done to minimise this resistance?
4. Does the outer cylinder have a function besides guiding the inner one?
5. What considerations affect the choice of how deeply submerged the cylinder needs to be? Explain.
6. What considerations affect the dimensions of the cylinders, in both length and diameter?
7. What material would be suitable for the walls of the cylinders?
8. What material would you suggest for the frame and the levers? Why?
9. Is a watertight seal between the two cylinders necessary? Why, or why not?
10. Which would suit human ergonomics better, to have a long, slow stroke, or a short, punchy one?
11. How high above the waterline would you estimate the centre of gravity of this craft to be, with a rider on board?
12. Would such a craft appear to be stable? Explain why or why not, in relation to the shift in position of the centre of gravity and the centres of buoyancy of the two hulls, when the craft tilts slightly.
13. Make a reasoned estimate of the force that a reasonably athletic operator could put into a pull-stroke, and of the length of a pull-stroke.
14. If this pull-stroke takes 2 seconds to carry out, and the forward stroke takes 1 second, while experiencing a resistance of 40 N, what would be the steady rate of energy expenditure by the operator?
15. If operated in still air on still water, estimate the value of the air resistance force on the craft at a speed of 5 m/s. *Tables of drag coefficients and how to use them are provided on pp164 - 166 of the author's Basic Engineering Mechanics Explained, vol. 3.*

Exercise 24.5   Hand-cranked grindstone

The grinding wheel of an old-fashioned hand-cranked (geared) grindstone consists of a stone of density 2700 kg/m³, diameter 788 mm and thickness 104 mm. The stone is brought to a speed of 60 r/min and then allowed to freewheel until it comes to rest, which takes 1 minute and 46 seconds.

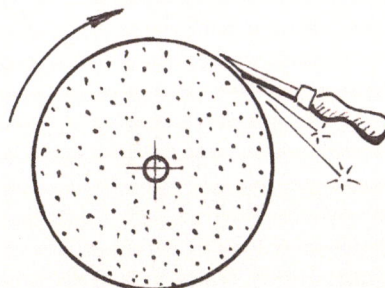

a. Determine the mass moment of inertia of the grindstone wheel. (Ignore the central hole and the axle through this hole: just consider it as a solid disc.)	
b. Determine the frictional torque in the bearings.	

Now, the stone is again accelerated, and is used to sharpen a tool, at a steady speed of 160 r/min. The tool is pressed against the surface of the stone with a radial force of 9 N. The coefficient of friction between tool and stone is 0.4.

c. If the person powering the grindstone stops turning the crank, but the tool is kept pressed against the stone with the same constant force, how long will the stone take to come to rest?	
d. What is the angular momentum of this stone at 160 r/min?	
e. What value of braking torque would be required to bring the stone to rest from 160 r/min, in 6 seconds? (Without the tool against the surface.)	
f. When turning at 160 r/min, what is the relative velocity of the stone's grinding surface, as seen from the tool tip?	
g. What steady power input is required from the person turning the crank while this tool is being sharpened?	

Answers: 10.63 kg.m² 0.6300 Nm 86.95 sec 178.1 kg.m²/s 29.05 Nm 6.601 m/s 34.32 W

Exercise 24.6

A thin rod, of mass 12 kg, length 1.8 m, and of uniform section and density,, is attached to a fixed point by a rope at its upper end. It is in equilibrium in the position shown. The lower end of the rod rests on a rubber mat on the floor. The coefficient of friction between the rod and the mat is 0.8.

Determine:

- The tension in the rope,
- The value of the friction force between the rod and the mat, and
- The upward reaction that the mat exerts on the rod.

For the answers to this exercise, see below exercise 24.7

Exercise 24.7

An L-shaped solid steel bar, with mass per unit length 8 kg/m, is fixed to a steel plate of thickness 5 mm by two steel rivets. Take the density of steel to be 7800kg/m³. Determine:

- The mass of the assembly, in kg, to 3 decimal places,
- The x-coordinate of the centre of gravity of the assembly (in mm, correct to one decimal place), and
- The y-coordinate of the centre of gravity of the assembly (same specifications for the answer as above)

The six answers to the above two exercises may be found among the following twelve jumbled values: 41.45 N; 26.29 N; 32.905 kg; 627.2 mm; 485.5 mm; 34.375 kg; 539.1 mm; 85.09 N; 28.954 kg; 32.74 N; 660.8 mm; 36.80 N

Exercise 24.8.a

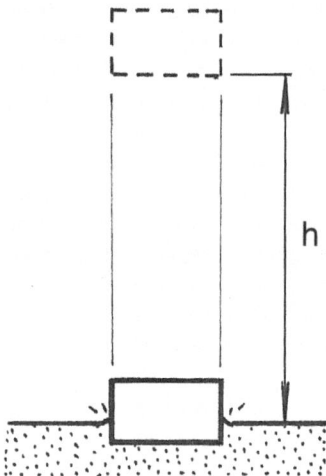

A concrete block of mass 2000 kg is dropped from a height of 20 m onto level soil. The block comes to rest, having made a hole 200 mm deep.

A negligible amount of soil splashes out of the hole.

Determine:

- The velocity of the block on striking the ground [19.809 m/s]
- The time during which the block decelerates [0.0219 seconds]
- The average deceleration of the block while penetrating the ground [981 m/s²]
- The average resisting force of the ground [1.962 MN]
- The work done to compress the soil [392.4 kJ]

Exercise 24.8.b

Sketch a velocity-time graph for the motion of the block in exercise 24.a, with the upward direction as positive. Show clearly the times taken for the two phases of motion of the block.

Solution:

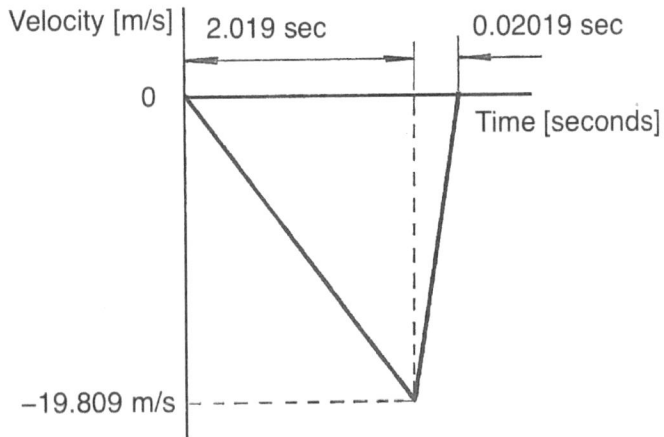

Exercise 24.8.c

No-one can know the answers to the following question without appropriate measurement. However, it would be interesting to see whether group participants' surmised values would agree broadly.

The kinetic energy that the block possessed before impact is distributed in several ways by the time it comes to rest. Estimate the percentage of this energy that repectively went into:

- Work done to compress the soil immediately under the block
- Shock waves sent through the surrounding ground
- Creating sound waves through the air
- Giving rise to heat, due to friction with the ground.

Appendix

Description of this author's series of three volumes 'Basic Engineering Mechanics Explained'.

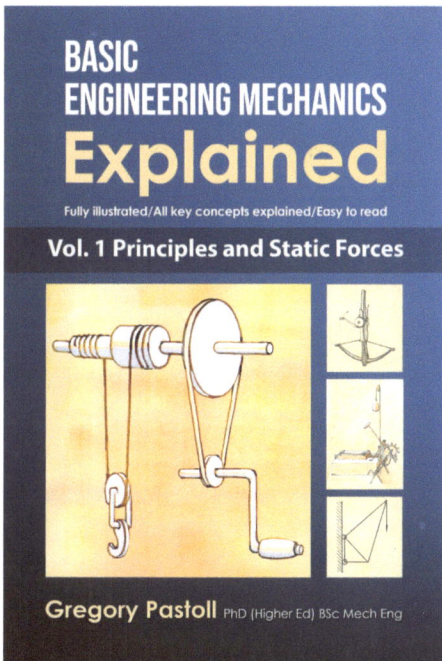

This series sets out in a reader-friendly way the essential principles of all the key concepts in basic mechanics and their application to a variety of situations in mechanical engineering.

These books would be found useful by any person either studying the subject, teaching it, or needing to apply it. That includes high school students intending to study engineering, first and second-year students at colleges and universities, and practising technicians and engineers.

One of the best endorsements of the value of these books came from an engineering lecturer whom the author has never met, who wrote in a review: *'I wish I had had these books at the start of my engineering career.'* A practising engineer from another country, whom the author has not met either, wrote: *'You did a wonderful job with the books. It's amazing how you always learn something new when you go over the basics. There are some topics which you have explained much better than the text books I worked with. Thank you for your work.'*

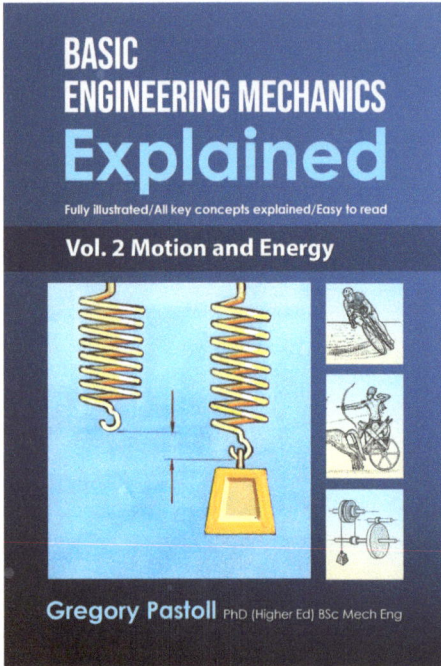

BASIC ENGINEERING MECHANICS
Explained
Fully illustrated/All key concepts explained/Easy to read
Vol. 2 Motion and Energy

Gregory Pastoll PhD (Higher Ed) BSc Mech Eng

The emphasis in these books is on approaching problems with common sense, and reasoning from first principles. The level of mathematics used is appropriately unpretentious, because the author believes that it is more important for a student of mechanics to understand mechanical principles than to engage in high-level mathematics and programming.

The basic principles of mechanics always remain constant, and if you understand them, you can grasp the essentials of any physical engineering situation.

The three volumes of 'Basic Engineering Mechanics Explained' between them contain 1000 illustrations and over 420 original exercises and examples. The solutions to many of the examples are shown with full workings. The answers to all questions requiring calculations are supplied.

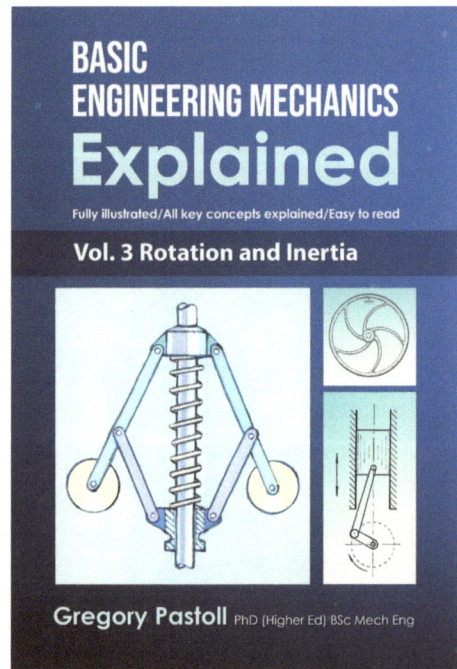

BASIC ENGINEERING MECHANICS
Explained
Fully illustrated/All key concepts explained/Easy to read
Vol. 3 Rotation and Inertia

Gregory Pastoll PhD (Higher Ed) BSc Mech Eng

Anyone who has mastered the principles that are set forth in that series will have an indisputably sound grasp of the basics of the science of mechanics, as applied to engineering. If you have at your disposal the problem-solving methods provided by that series, you will be able to solve all the exercises in the present book, with confidence.

Contents outline for the series 'Basic Engineering Mechanics Explained'

Volume 1: Principles and static forces

Volume 2: Motion and energy

Volume 3: Rotation and inertia

About the author

Gregory Pastoll taught basic Engineering Mechanics to first- and second- semester students in a polytechnic/university of technology environment in Cape Town for a total of 14 years. For much of this time he was course coordinator and examiner in Mechanics 1, and later in Mechanics 2.

When teaching, he experimented widely with ways of getting students to engage with the subject and to be motivated by interest in the challenges he set for them.

He estimates that over 4000 students have attended his classes.

He was also a consultant on university teaching methods at the University of Cape Town for a further 14 years.

He obtained a BSc Mech. Eng. from the University of the Witwatersrand in 1973, an M.Phil from the University of Cape Town in 1986, and a PhD in Higher Education from the same university in 1994.

Besides his books on engineering, the author has written books on education and some humorous fiction. All his books can be obtained from online booksellers.

Answers to the true/false tests

The table below indicates the numbering of the T/F tests that go with each respective topic in this book. The subsequent tables list the correct answers for all these tests.

Set	Topic	T/F tests
1	Concepts, quantities, principles & laws	1a 1b
2	Expressing numbers in engineering	2a 2b
3	Forces, components, equilibrium of particles	3a 3b
4	Force moments, torque, equilib. rigid bodies	4a
5	Centres of mass, of gravity & centroids	5a
6	Forces in structures	6a 6b
7	Friction (dry, sliding)	7a 7b
8	Buoyancy	8a 8b
9	Linear motion with uniform acceleration	9a 9b
10	Motion influenced by gravity	10a
11	Rotary motion with uniform acceleration	11a 11b
12	Work, energy and power	12a 12b 12c
13	Simple lifting machines	13a 13b 13 c
14	Linearly accelerating systems	14a
15	Linear momentum	15a 15b
16	Relative velocity	16a 16b
17	Centripetal and centrifugal force	17a 17b
18	Rotational inertia	18a 18b
19	Rotational and linear inertia combined	19a
20	Kinetic energy of rotation; angular momentum	20a 20b
21	Simple harmonic motion	21a 21b
22	Basic vehicle dynamics	22a 22b
23	Exercises combining several topics	23a; b; c; d

In the tables below, a T marks those statements in each respective test that are true. The remainder are false.

Test	Q1	Q2	Q3	Q4	Q5	Q6	Q7	Q8	Q9	Q10
1a		T	T	T	T		T		T	
1b			T			T		T	T	T
2a	T	T		T	T	T			T	
2b		T	T	T	T	T	T			
3a	T	T	T	T						T
3b		T		T	T		T			T
4a	T	T			T	T		T		
5a	T		T	T			T	T		T
6a		T	T		T		T	T		T
6b	T		T		T		T	T	T	
7a		T	T		T	T		T	T	
7b		T			T	T		T	T	
8a			T	T	T			T	T	T
8b	T		T				T			T
9a		T	T		T			T		T
9b	T		T	T	T	T			T	T
10a	T		T	T			T			
11a		T		T		T	T		T	T
11b	T	T				T	T	T	T	
12a				T	T			T	T	T
12b			T	T		T	T	T	T	
12c	T		T	T	T		T	T	T	

Test	Q1	Q2	Q3	Q4	Q5	Q6	Q7	Q8	Q9	Q10
13a	T	T	T	T		T			T	
13b	T		T		T	T	T		T	T
13c	T	T			T					
14a	T				T		T		T	
15a		T	T		T	T	T		T	
15b	T		T	T		T		T		T
16a		T		T	T	T	T	T		T
16b	T		T	T	T		T	T		T
17a	T	T			T		T	T	T	T
17b	T		T	T		T	T		T	T
18a		T	T	T		T		T		
18b	T						T	T	T	T
19a		T				T		T	T	T
20a	T		T	T	T		T			T
20b		T	T	T		T	T		T	
21a		T	T	T		T		T	T	
21b		T	T	T			T	T		
22a	T		T		T	T		T		
22b		T	T			T	T			T
23a		T		T		T	T	T		
23b	T			T	T	T	T		T	T
23c			T	T	T		T	T		T
23d			T	T				T		

www.ingramcontent.com/pod-product-compliance
Lightning Source LLC
Chambersburg PA
CBHW050906210326
41597CB00002B/45